Ben Stacy Jerrik (Ed.)

Nebraska Highway 45

Ben Stacy Jerrik (Ed.)

Nebraska Highway 45

Nebraska Highway 32, U.S. Route 275, Newman Grove, Nebraska, Nebraska Highway 91

Part Press

Publisher:
Part Press is a trademark of
International Book Market Service Ltd., 17 Rue Meldrum, Beau Bassin, 1713-01 Mauritius
Email: info@bookmarketservice.com
Website: www.bookmarketservice.com

Published in 2013

Printed in: U.S.A., U.K., Germany. This book was not produced in Mauritius.

ISBN: 978-613-9-21659-8

Contents

Articles

References

T_(New_York_City_Subway_service)

Second Avenue Local	
Northern end	125th Street
Southern end	Hanover Square
Stations	16

The **T Second Avenue Local** is a future rapid transit service of the New York City Subway. It will be colored turquoise on route signs, station signs, and the official subway map, since it will use the IND Second Avenue Line through Manhattan.

The **T** service will operate at all times between 125th Street in Harlem, Manhattan, and Hanover Square in Financial District, Manhattan via Second Avenue in Manhattan.

History

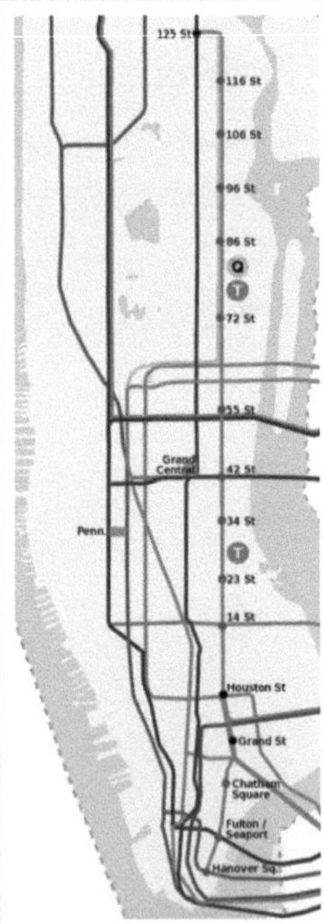

Map of the future Second Avenue Subway (SAS), with the T service shown in light-blue.

T was previously used for a service on the BMT West End Line, originally **BMT 3**. It was discontinued in 1967, but a shuttle, designated **TT**, continued into 1968,[1] when it was replaced by the **B** train via Sixth Avenue in Manhattan. With the closing of the Manhattan Bridge's north side in 2001, the **B** in Brooklyn was replaced by the **W** via Broadway. In 2004, the Manhattan Bridge reopened for subway service and the **W** was replaced by the D via Sixth Avenue.[2]

T currently appears on R32 rollsigns as a black letter on a white circle.

Station Listing

Station	Phase	Transfers & Notes
125th Street	2	Northern terminal station for **Q** train (Phase 2) and **T** train (Phase 3) 4 5 6 <6> trains (IRT Lexington Avenue Line) M60 bus to LaGuardia Airport connection to Harlem – 125th Street (Metro-North Railroad) at Lexington Avenue and 125th Street
Provision for expansion to The Bronx		
116th Street	2	
106th Street	2	
96th Street	1	Northern terminal station for **Q** train in Phase 1
86th Street	1	
72nd Street	1	
Q train splits to BMT Broadway Line via BMT 63rd Street Line (Phase 1)		
T train continues down Second Avenue (Phase 3)		
55th Street	3	E M trains (IND Queens Boulevard Line) at Lexington Avenue – 53rd Street 4 6 <6> trains (IRT Lexington Avenue Line) at 51st Street
42nd Street	3	7 <7> trains (IRT Flushing Line) S train (IRT 42nd Street Shuttle) 4 5 6 <6> trains (IRT Lexington Avenue Line) at Grand Central – 42nd Street connection to Grand Central Terminal (Metro-North Railroad & Long Island Rail Road once East Side Access Project is completed.)
34th Street	3	
23rd Street	3	
14th Street	3	L train (BMT Canarsie Line) at Third Avenue
Houston Street	3	F train (IND Sixth Avenue Line) at Second Avenue Southern terminal station for **T** train (Phase 3)
Grand Street	4	B D trains (IND Sixth Avenue Line)
Chatham Square	4	at Worth Street
Seaport	4	at Fulton Street
Hanover Square	4	at Old Slip Southern terminal station for **T** train (Phase 4)
Provision for expansion to Brooklyn		

See also

- Second Avenue Subway

External links

- Map of the Proposed "T" Subway Service [3]

References

[1] http://www.nycsubway.org/perl/show?36091
[2] BMT West End Line (http://www.nycsubway.org/lines/westend.html)
[3] http://www.mta.info/capconstr/sas/images/sas_map_lg.gif

New_York_City_Subway

New York City Subway

New York City Subway

Top: A number **4** train made up of R142 cars enters the Fordham Road station.
Bottom: An **E** train made up of R160A cars waits for passengers at the 42nd Street – Port Authority Bus Terminal station.

Background	
Locale	New York City
Transit type	Rapid transit
Number of lines	34 physical lines (1 under construction) 24 services (1 planned)
Number of stations	468 (the MTA classification) 421 (when compared to international standards) 4 under construction

Daily ridership	5,284,295 (weekdays, 2011) 3,033,660 (Saturdays, 2011) 2,367,261 (Sundays, 2011)[1]
Website	www.mta.info/nyct [2]
Operation	
Began operation	October 27, 1904 (first section of subway) July 3, 1868 (first elevated operation) October 9, 1863 (first railroad operation)[3]
Operator(s)	New York City Transit Authority (NYCTA)
Technical	
System length	mi (337 km) (route length) 656 mi (km) (track length, revenue) 842 mi (km) (track length, total)
Track gauge	4 ft 8 $^1/_2$ in (1435 mm) Standard gauge
Electrification	625V (DC) third rail[4]

The **New York City Subway** is a rapid transit system owned by the City of New York in the United States and leased to the New York City Transit Authority,[5] a subsidiary agency of the Metropolitan Transportation Authority. It is the most extensive public transportation system in the world by number of stations, with 468 stations in operation (421, if stations connected by transfers are counted as single stations).[1] The New York City Subway is also one of the world's oldest public transit systems. Overall, the system contains mi (337 km) of routes,[6] translating into 656 miles (km) of revenue track; and a total of 842 miles (km) including non-revenue trackage.[7] In 2011, the subway delivered over 1.64 billion rides,[8] averaging approximately 5.3 million rides on weekdays, about 3.0 million rides on Saturdays, and about 2.4 million rides on Sundays.[1]

By annual ridership, the New York City Subway is the seventh busiest rapid transit rail system in the world; the metro (subway) systems in Tokyo, Seoul, Moscow, Beijing, Shanghai, and Guangzhou record a higher annual ridership.[9]

In the United States, four rapid transit systems offer rail service 24 hours per day and 365 (or 366) days per year: the New York City Subway, PATH trains, PATCO trains, and parts of the Chicago 'L' system.

Overview

Stations are located throughout the boroughs of Manhattan, Brooklyn, Queens and the Bronx. While Staten Island does have a rail line, the Staten Island Railway, it is not officially considered part of the subway. In addition, the Staten Island Railway does not connect to any other trains, so any passengers wishing to visit another borough must take a ferry or bus.

Almost all routes pass through Manhattan (the exceptions being the G, Franklin Avenue Shuttle and Rockaway Park Shuttle trains), and all but five of the 468 stations are served 24 hours a day.[10] Contrary to its name, the New York City Subway system is not entirely underground. Large portions of the system (especially outside of Manhattan) are elevated, on embankments, or in open cuts, and a few stretches of track run at ground level.

An entrance to the Times Square – 42nd Street station, the busiest station of the New York City Subway.[11]

Many lines and stations have both express and local services. These lines have three or four tracks. Normally, the outer two are used for local trains while the inner one or two are used for express trains. Stations served by express trains are typically major transfer points or destinations. The BMT Jamaica Line uses skip-stop service on portions, where two services (J and Z trains) operate over the line during rush hours in the peak direction and many stations are only served by one of the two.

In 2005, the New York City Subway hit a 50-year record in usage with a ridership of 1.45 billion.[12] The trend toward higher ridership continued into 2008; MTA released figures that subway use was up 6.8 percent for January and February as higher gasoline prices encouraged riders to use mass transit over automobiles.[13]

The Citizens Budget Commission reported in 2011 that the subway system was among the most cost-efficient in the country, with the lowest cost per passenger trip and second-lowest cost per passenger mile.[14]

According to the United States Department of Energy, energy expenditure on the New York City Subway rail service was 3492 BTU/passenger mile (2289 kJ/passenger km) in 1995. This compares with 3702 BTU/passenger mile (2427 kJ/passenger km) for automobile travel.[15] However, the figure for automobiles is averaged over the entire United States. Driving a car in New York City is significantly less efficient due to the highly urbanized environment.[16]

History

Political cartoon critical of IRT service in 1905. The IRT is labeled as the "Interborough Rattled Transit".

	External videos
	Subway Colors and Names [17], Metropolitan Transportation Authority; July 25, 2010; 2:09 YouTube video clip
	Ten Years Later: MTA Reflects on 9/11 [18], Metropolitan Transportation Authority; September 7, 2011; 4:21 YouTube video clip

A demonstration for an underground transit system in New York City was first built by Alfred Ely Beach in 1869. His Beach Pneumatic Transit only extended 312 feet (m) under Broadway in Lower Manhattan and exhibited his idea for a subway propelled by pneumatic tube technology. The tunnel was never extended for political and financial reasons, although extensions had been planned to take the tunnel southward to The Battery and northwards towards the Harlem River.[19] The Beach subway was demolished when the BMT Broadway Line was built in the 1910s; thus, it was not integrated into the New York City Subway system.

The first underground line of the subway opened on October 27, 1904, almost 35 years after the opening of the first elevated line in New York City, which became the IRT Ninth Avenue Line. The Great Blizzard of 1888 helped demonstrate the benefits of an underground transportation system. The oldest structure still in use opened in 1885 as part of the BMT Lexington Avenue Line in Brooklyn and is now part of the BMT Jamaica Line. The oldest right-of-way, that of the BMT West End Line, was in use in 1863 as a steam railroad called the Brooklyn, Bath and Coney Island Rail Road.

Map of the 1906 IRT system in Manhattan

By the time the first subway opened, the lines had been consolidated into two privately owned systems, the Brooklyn Rapid Transit Company (BRT, later Brooklyn–Manhattan Transit Corporation, BMT) and the Interborough Rapid Transit Company (IRT). The city was closely involved: all lines built for the IRT and most other lines built or improved for the BRT after 1913 were built by the city and leased to the companies. The first line of the city-owned and operated Independent Subway System (IND) opened in 1932; this system was intended to compete with the private systems and allow some of the elevated railways to be torn down, but kept within the core of the City due to the low amount of startup capital provided to the municipal Board of Transportation, the later MTA, by the state.[5] This required it to be run 'at cost', necessitating fares up to double the five cent fare popular at the time.[20]

In 1940, the two private systems were bought by the city and some elevated lines closed immediately while others closed soon after. Integration was slow, but several connections were built between the IND and BMT, and now operate as one division called the B Division. Since the IRT tunnel segments are too small and stations too narrow to accommodate B Division cars, and contain curves too sharp for B Division cars, the IRT remains its own division, A Division.

The New York City Transit Authority, a public authority presided by New York City, was created in 1953 to take over subway, bus, and streetcar operations from the city, and placed under control of the state-level Metropolitan Transportation Authority in 1968.

In 1934, transit workers of the BRT, IRT, and IND founded the Transport Workers Union of America, organized as Local 100. Local 100 remains the largest and most influential local of the labor union. Since the union's founding, there have been three union strikes over contract disputes with the MTA, 12 days in 1966, 11 days in 1980, and three days in 2005.

The Cortlandt Street station after the World Trade Center collapse

The September 11th attacks resulted in service disruptions on lines running through Lower Manhattan, particularly the IRT Broadway – Seventh Avenue Line, which ran directly underneath the World Trade Center between the Chambers Street and Rector Street stations. Sections of the tunnel, as well as the Cortlandt Street station, which was directly underneath the Twin Towers, were severely damaged by the collapse and had to be rebuilt, requiring suspension of service on that line south of Chambers Street. Ten other nearby stations were closed while dust and debris were cleaned up. By March 2002, seven of those stations had reopened. The rest (except for Cortlandt Street on the IRT Broadway – Seventh Avenue Line) reopened on September 15, 2002 along with service south of Chambers Street.[21] [22]

Construction methods

42nd Street station improvements, 1987.

When the IRT subway debuted in 1904, the typical tunnel construction method was cut-and-cover. The street was torn up to dig the tunnel below before being rebuilt from above. This method worked well for digging soft dirt and gravel near the street surface. However, mining shields were required for deeper sections, such as the Harlem and East River tunnels, which used cast-iron tubes, segments between 33rd and 42nd streets under Park Avenue, 116th Street and 120th Street under Broadway, and 145th Street and Dyckman Street (Fort George) under Broadway and Saint Nicholas Avenue as well as the tunnel from 96th Street to Central Park North - 110th Street & Lenox Avenue, all of which used either rock or concrete-lined tunnels.[23]

About 40% of the subway system runs on surface or elevated tracks, including steel or cast iron elevated structures, concrete viaducts, embankments, open cuts and surface routes. All of these construction methods are completely grade-separated from road and pedestrian crossings, and most crossings of two subway tracks are grade-separated with flying junctions.

Lines and routes

125th Street station on the IRT Broadway – Seventh Avenue Line

Many rapid transit systems run relatively static routings, so that a train "line" is more or less synonymous with a train "route". In New York, routings change often as new connections are opened or service patterns change. Within the nomenclature of the subway, the "line" describes the physical railroad track or series of tracks that a train "route" uses on its way from one terminal to another. "Routes" (also called "services") are distinguished by a letter or a number and "Lines" have names. They are also designations for trains, as exemplified in the Billy Strayhorn song Take the "A" Train. This terminology is also used to a loose extent in the Taipei Metro.

A digital sign on the side of an R142 **4** train

There are 24 train services in the subway system, including three short shuttles. Each route has a color and a local or express designation representing the Manhattan trunk line of the particular service. The color lime green is exclusively assigned to the Crosstown Line route since it operates entirely outside Manhattan while the shuttles are all assigned dark slate gray.[24] The 1, 6, 7, C, G, L, M and R trains are fully local; making all stops. The 2, 3, 4, 5, <6>, <7>, A, B, D, E, F, N and Q trains have portions of express and local service. The J train normally operates local, but during rush hours it is joined by the Z train in the peak direction. Both run local, express or skip-stop on different parts of their route. The letter S is used for three shuttle services. The previous designation H has been put into use for a special fare-free shuttle service along part of the Rockaway Shuttle S and A routes in the aftermath of Hurricane Sandy in 2012.[25]

A Division (IRT) consists of:

Route	Name
1	Broadway – Seventh Avenue Local
2	Seventh Avenue Express
	Seventh Avenue Express
4	Lexington Avenue Express
5	Lexington Avenue Express
6	Lexington Avenue Local (Pelham Local/Express)
7	Flushing Local/Express
S	42nd Street Shuttle

B Division (BMT/IND) consists of:

Route	Name	Route	Name
A	Eighth Avenue Express	L	14 Street – Canarsie Local
B	Sixth Avenue Express	M	Sixth Avenue Local
C	Eighth Avenue Local	N	Broadway Local
D	Sixth Avenue Express	Q	Broadway Express
E	Eighth Avenue Local	R	Broadway Local
F	Sixth Avenue Local	S	Franklin Avenue Shuttle
G	Crosstown Local	S H	Rockaway Park Shuttle
J	Nassau Street Local		Nassau Street Express

The current color system depicted on official subway maps was proposed by R. Raleigh D'Adamo, a lawyer who entered a contest sponsored by the Transit Authority in 1964. D'Adamo proposed replacing a map that used only three colors (representing the three operating entities of the subway network) with a map that used a different color for each service. D'Adamo's contest entry shared first place with two others and led to the Transit Authority adopting a multi-colored scheme.[26] However, the lines and services are not referred to by color (e.g., Blue Line or Green Line), although the colors are often assigned through their groups.

Though the subway system operates on a 24-hour basis, some of the designated routes do not run, run as a shorter route or run with a different stopping pattern during late night hours. In addition to these regularly scheduled changes, because there is no nightly system shutdown for maintenance, tracks and stations must be maintained while the system is operating. To accommodate such work, services are usually changed during midday, overnight hours, and weekends.[27]

Subway map

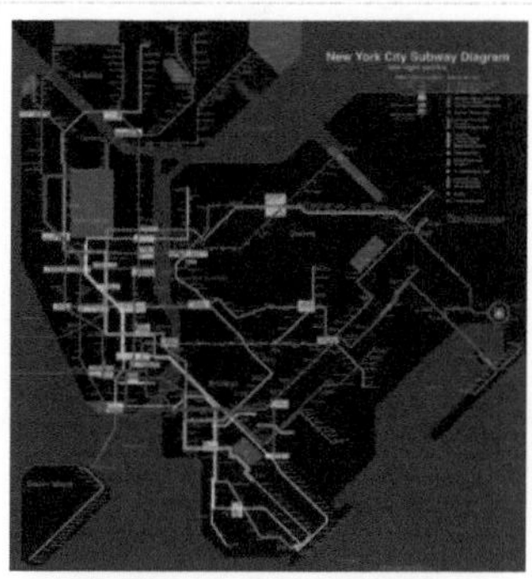

Late night subway service map

The current official transit maps of the New York City Subway are based on a 1979 design by Michael Hertz Associates. The maps are not geographically accurate due to the complexity of the system (i.e. Manhattan being the smallest borough, but having the most services), but are known to help tourists navigate the city, as major city streets are shown alongside the subway stations serving them. The newest edition of the subway map, which took effect on June 27, 2010, reflects the latest service changes and also makes Manhattan bigger and Staten Island smaller.[28] [29] A late night-only version of the map was introduced on January 30, 2012.[30]

Part of the reason for the current incarnation is that earlier diagrams of the subway (the first being produced in 1958), while being more aesthetically pleasing, had the perception of being more geographically inaccurate than the diagrams today. The design of the subway map by Massimo Vignelli, published by the MTA between 1972 and 1979, has since become recognized in design circles as a modern classic; however, the MTA deemed the map flawed due to its placement of geographical elements.[31] [32]

On September 16, 2011, the MTA introduced a Vignelli-style interactive subway map, aptly called "The Weekender",[33] to its website.[34] As the title suggests,[35] it is a way for riders to get information about any planned work, from late Friday night to early Monday morning, that is going on either on a service(s) or station(s) of the subway during the weekend only.[36] [37] On June 11, 2012, the MTA duplicated "The Weekender" site as a free mobile app download for the iPhone, iPod Touch, and the iPad. An Android version is still being developed.[38] [39]

There are several privately produced schematics which are available online or in published form, such as those by Hagstrom Map.[40] Additionally, the New York City subway map has served as the subject of artistic endeavors. Among these are works by Fadeout Design and by Alexander Chen.[41] [42]

Stations facilities and amenities

Entrance to Broad Street station with its red lamps

Lit-up exit sign on a turnstile at 63rd Drive – Rego Park

The IND Eighth Avenue Line station at 59th Street – Columbus Circle

Station and concourse

Many stations have mezzanines. These allow for passengers to enter from multiple entrances and proceed to the correct platform without having to cross the street before entering. They also allow for crossover between the uptown and downtown platforms.

Passengers enter a subway station through stairs towards station booths and vending machines to buy their fare, which is currently stored in a MetroCard. After swiping the card at a turnstile, customers continue to the platforms. Some subway lines in northern Manhattan and the other boroughs have elevated tracks to which passengers climb up to the platforms and station houses via stairs, escalators, or elevators.

Globe lamps

At most of the system's entrances and exits sits a lamp post or two bearing a colored spherical lamp. These lights were once used to indicate the station's availability. A green lamp means that the station is open and running 24 hours a day, while a red lamp means that it is an exit-only or part-time. A yellow lamp once indicated a part-time station, but in an attempt to simplify the system, was eventually phased out and replaced by red lamps, now indicating both a part-time entrance and exit-only. With the introduction of the MetroCard in 1994, some station entrance/exit configurations have been changed by merging some entrances and exits, thus making this light system no longer accurate.[43]

Platforms

A typical subway station has waiting platforms ranging from 500 to 600 feet (to m) long. Due to the large number of transit lines, one platform or set of platforms often serve more than one service. Passengers need to look at the overhead signs at the platform entrance steps and over each track to see which trains stop there and when, and at the arriving train to see which one it is.

There are a number of common platform configurations:

- On a double track line, a station may have one center island platform used for trains in both directions, or 2 side platforms, one for a train in each direction.

- For lines with three or four tracks with express service, local stops will have side platforms and the middle one or two tracks will not stop at the station. On these lines, express stations have two island platforms, one for the local and express in one direction, and another for the local and express in the other direction. Each island platform provides a cross-platform interchange between local and express services. Almost everywhere expresses run, they run on the inner one (of 3) or two (of 4) tracks and locals run on the outer two tracks. In a 3-track configuration, the center express track can be used toward the center of the city in the morning and away from the center in the afternoon and evening, though not every 3-track line has that express service.

Three four-track express stations have an island platform for the center express tracks and two side platforms for the outside local tracks. These three stations are connected to major railway stations and the next station along the line is also an express station with the more common platform configuration. The purpose of splitting the platforms is to limit overcrowding by preventing cross-platform interchanges between local and express services. This occurs at Atlantic Avenue – Barclays Center on the IRT Eastern Parkway Line (2 3 4 5 trains) with the adjacent express station Nevins Street, where the connection is to the Atlantic Terminal of the Long Island Rail Road; and 34th Street – Penn Station on both the IRT Broadway – Seventh Avenue Line (1 2 3 trains) and IND Eighth Avenue Line (A C E trains), with adjacent express stations at Times Square – 42nd Street and 42nd Street – Port Authority Bus Terminal, where a connection is available to Pennsylvania Station, one of the two major Manhattan train stations. This does not occur with the connection to New York's other major station, Grand Central Terminal, at Grand Central on the IRT Lexington Avenue Line (4 5 6 <6> trains), which has no adjacent express station.

There is one notable 6-track local station, DeKalb Avenue, where trains to or from the Manhattan Bridge (B D Q) either stop at the outer tracks of one of the island platforms, or pass through and bypass the station on the middle tracks ("express tracks") (D N). Trains to or from the Montague Street Tunnel (N R) stop across the platform from the respective outer track.

Artwork

Many stations are decorated with intricate ceramic tile work, some of it dating back to 1904 when the subway first opened. The subway tile artwork tradition continues today.

The "Arts for Transit" program oversees art in the subway system.[44] Permanent installations, such as sculpture, mosaics, and murals; photographs displayed in lightboxes encourage people to use mass transit.[45] [46] In addition, commissioned art displayed in stations and "art cards", some displaying poetry, are in many of the trains themselves in unused advertisement fixture slots. Some of the art is by internationally known artists such as Elizabeth Murray's *Blooming*, displayed at Lexington Avenue / 59th Street station.[47]

Accessibility

The Crown Heights – Utica Avenue station is one of a group of stations that became accessible after station reconstruction

Since the majority of the system was built before 1990, the year the Americans with Disabilities Act (ADA) went into effect, many New York City Subway stations were not designed to be handicapped-accessible. Since then, elevators have been built in newly constructed stations to comply with the ADA. (Most grade-level stations required little modification to meet ADA standards.) In addition, the MTA identified "key stations," high-traffic and/or geographically important stations, which must conform to the ADA when they are extensively renovated.[48] As of June 2011, there are 89 currently accessible stations; many of them have AutoGate access.[49] [50] [51]

Street elevator serving as an entrance to the 66th Street – Lincoln Center station

Entertainment

A typical scene of musicians performing on the platform of the Broadway – Lafayette St station.

Since 1987, MTA has sponsored the "Music Under New York" (MUNY) program[52] in which street musicians enter a competitive contest to be assigned to the preferred high traffic locations. Each year, applications are reviewed and approximately 70 eligible performers are selected and contacted to participate in live auditions held for one day.[53] [54] [55] [56] [57]

At present, more than 100 soloists and groups participate in MUNY providing over 150 weekly performances at 25 locations throughout the transit system. For example, Mr. M. Salieu Suso, a kora player from The Gambia, plays at Union Square.[58]

In addition, any musician/entertainer may perform in subway mezzanines and platforms. On platforms, there may be no amplifications as this is part of MTA policies:[59]

> "The New York City Transit (NYCT) is a subdivision of the Metropolitan Transportation Authority (MTA) that operates the city's subways and buses. The NYCT authorizes these types of free expression in subway stations: "Public speaking; distribution of written materials; solicitation for charitable, religious or political causes; and artist performances, including the acceptance of donations."

Performers must not be within 25 feet (m) of a token booth or 50 feet (m) from a MTA office/tower, blocking access to an escalator, stairwell, or elevator, interfering with transit services or passenger movement; or in an area where construction is occurring. In addition, performance is prohibited during public service announcements and may be no louder than 85 dBA at 5 feet (m) away or 70 dBa at 2 feet (m) from a token booth. Performances are prohibited in subway cars.

Restrooms

Restrooms at Church Avenue (IND Culver Line)

Restrooms are rare in the subway system as only 129 open restrooms are in 77 of the system's 468 stations.[60] Most station restrooms previously open to the public have been closed to the public and converted to storage spaces or for employee use only. However, there are a few major stations that have operating restrooms, including on the concourse of 42nd Street – Port Authority Bus Terminal, Chambers Street, 57th Street - Seventh Avenue, Coney Island – Stillwell Avenue, and Lexington Avenue / 59th Street.[61]

Retail

Former women's restroom converted into newsstand at Astor Place (IRT Lexington Avenue Line)

Some platforms have newspaper stands that sell various items including newspapers and food. The MTA also installed retail spaces within paid areas in selected stations, including the station concourses of the Times Square complex and the Sixth Avenue concourse at 42nd Street – Bryant Park.

According to the MTA, the New York City Subway is home to 345 retail spaces, making over US$70 million in rent and licensing fees in 2009 for the authority. It is continuing to make efforts in attracting more diverse retailers and vendors to set up shop in the subway system.[62]

Connections

Rapid transit and rail connections are available at designated stations to Amtrak, Long Island Rail Road, AirTrain JFK, Metro-North Railroad, New Jersey Transit and PATH. Connections to the Staten Island Ferry and privately operated ferries such as NY Waterway and New York Water Taxi, as well as intercity and commuter bus lines at the Port Authority Bus Terminal and George Washington Bridge Bus Terminal, are also available. Free MetroCard-only transfers to buses are available to MTA New York City Transit buses (including the bus rapid transit Select Bus Service), MTA Bus Company, NICE buses (Nassau County) and Bee-Line buses (Westchester County).

Rolling stock

R46 R train in Lower Manhattan

As of June 2011, the New York City Subway has 6,292 cars on the roster.[63] A typical New York City Subway train consists of 8 to 11 cars, although shuttles can have as few as two, and the train can range from 150 to 600 feet (to m) in length. Trains not in revenue service can be up to 12 cars long.

A *Vaktrak* track vacuuming train

Driver's cab of an R160B series car on the N train

The system maintains two separate fleets of cars, one for the IRT lines and another for the BMT/IND lines. All BMT/IND equipment is about 10 feet (3.05 m) wide and either 60 feet 6 inches (m) or 75 feet (22.86 m) long whereas IRT equipment is approximately 8 feet 9 inches (m) wide and 51 feet 4 inches (m) long. There is also a special fleet of BMT/IND cars that is used for operation in the BMT Eastern Division, consisting of R42 married pairs and R143 and R160A four-car sets. Due to clearance issues on various sharp curves, 75-foot (22.86 m) long cars are not permitted on BMT Eastern Division trackage.

Cars purchased by the City of New York since the inception of the IND and the other divisions beginning in 1948 are identified by the letter "R" followed by a number; e.g.: R32. This number is the contract number under which the cars were purchased. Cars with nearby contract numbers (e.g.: R1 through R9, or R26 through R29, or R143 through R160B) may be relatively identical, despite being purchased under different contracts and possibly built by different manufacturers.

The MTA has been incorporating newer subway cars into its stock in the past decade. Since 1999, the R142, R142A, R143, R160A, and R160B fleet have been added into service.[64] [65] These cars feature recorded announcements for station information, closing doors, and other general messages in lieu of conductor announcements, although live conductor announcements can still be made. The recordings began in the late 1990s and featured Bloomberg Radio on-air speakers, who volunteered at the request of their employer and future city mayor Michael Bloomberg. Voices include Jessica Ettinger Gottesman, Diane Thompson, Charlie Pellett, and Catherine Cowdery. With regards to why certain messages are voiced by males and others by females, MTA spokesperson Gene Sansone said in 2006 that, "Most of the orders are given by a male voice, while informational messages come from females. Even though this happened by accident, it is a lucky thing because a lot of psychologists agree that people are more receptive to orders from men and information from women".[66]

On March 24, 2012, the MTA announced that it ordered 300 R179 subway cars from Bombardier.[67] [68] The total price of the contract is US$600 million, with the first test train arriving in late 2014.

Fares

An NYCTA token from the mid-20th century

Token and change

From the inauguration of IRT subway services in 1904[69] until the unified system of 1948 (including predecessor BMT and IND subway services), the fare for a ride on the subway of any length was 5 cents (nickel). On July 1, 1948, the fare was increased to 10 cents (dime), and since then has steadily risen. When the New York City Transit Authority was created in July 1953, the fare was raised to 15 cents and a token issued. Until April 13, 2003, riders paid the fare with tokens purchased from a station attendant. The tokens were changed periodically as prices changed. For the 75th anniversary of the subway in 1979 (also called the Diamond Jubilee), a special token with a small off-center diamond cutout and engraved images of a 1904 subway car and kiosk were issued. Many were purchased for keepsakes and were not used for rides. The last iteration of tokens featured a hole in the middle and phased out in 2003 when the fare rose to $2.[70]

Token sucking

It was a common scam to circumvent the payment of fares by jamming the token slot in an entrance gate with paper. A passenger would insert a token into the turnstile, be frustrated when it did not open the gate, and have to spend another token to enter at another gate. A token thief would then suck the token from the jammed slot with their mouth. This could be repeated many times as long as no police officers spotted the activity. Some token booth attendants sprinkled chili powder in the slots to discourage "token sucking".[71] Token sucking (also known as stuff 'n' suck) was charged under theft of services, criminal tampering and criminal mischief.[72]

Token War with Connecticut

Transit riders discovered in the early 1980s that tokens purchased for use in the Connecticut Turnpike toll booths were of the same size and weight as New York City subway tokens. Since they cost less than one third as much, they began showing up in subway collection boxes regularly.[73] Connecticut authorities initially agreed to change the size of their tokens,[74] but later reneged and the problem went unsolved until 1985, when Connecticut discontinued the tolls on its turnpike.[75] At that time, the MTA was paid 17.5 cents for each of more than two million tokens that had been collected during the three-year "token war."[75]

MetroCard

The current Metrocard design

In 1994, the subway system introduced a fare system called the MetroCard, which allows riders to use cards that store the value equal to the amount paid to a station booth clerk or vending machine. The MetroCard was enhanced in 1997 to allow passengers to make free transfers between subways and buses within two hours; several MetroCard-only transfers between subway stations were added in 2001. With the addition of unlimited-ride MetroCards in 1998 (for 7-day and 30-day periods, later 1-day "Fun Pass" and 14-day periods, both of which have been discontinued), the New York City Transit system was the last major transit system in the United States with the exception of BART in San Francisco to introduce passes for unlimited bus and rapid transit travel.[76]

Renovation and expansion plans

Current expansion projects include the:

- Second Avenue Subway on the Upper East Side of Manhattan
- 7 Subway Extension to the west side of Manhattan
- Fulton Center in Lower Manhattan[77] [78] [79] [80] [81] [82]

In August 2006, the MTA revealed that all future subway stations, including ones built for the Second Avenue Subway and Flushing Line extension, will have platforms outfitted with air-cooling systems.

Operations

Pending legislation would merge the subway operations of MTA New York City Transit with Staten Island Railway to form a single entity called MTA Subways.[83] The Staten Island Railway operates with R44 subway cars on a fully grade-separated right-of-way, but is typically not considered part of the subway, and is connected only via the free, city-operated Staten Island Ferry.

In the early 21st century, plans resurfaced for a major expansion, the Second Avenue Subway. This line had been planned in the 1920s, but was delayed several times since. Construction was started in the 1970s, but discontinued due to the city's fiscal crisis. Some small portions remain intact in Chinatown, the East Village, and the Upper East Side, but they are each quite short and thus remain unused.[84]

FASTRACK on the IND Eighth Avenue Line

FASTRACK

In 2012, the MTA introduced a new maintenance program, FASTRACK, to speed up repair work. This program involves a more drastic approach than previous construction, and shuts down an entire trunk line in Manhattan for four consecutive weeknights.[85] According to the MTA, this new program proved much more efficient and quick than regular service changes, especially because it happened at night and not the weekend, when most transit closures had occurred before.[86]

Technology

Train arrival times

New train arrival signs on the BMT Canarsie Line

In 2003, the MTA signed a $160 million contract with Siemens Transportation Systems to install digital next-train arrival message boards, called Public Address/Customer Information Screens (PA/CIS) at 158 of its IRT stations.[87] These signs were to be different from the current LED signs that display the current date and time. However, many problems arose with the software used in Siemens programming, and the MTA stopped payment to the company in May 2006.[88] The MTA threatened to drop Siemens, but about a month later Siemens announced they had fixed the problem. The signs were scheduled to begin operation in late 2007.[89]

A different system was eventually developed, tested, and installed successfully on the L train.[90] In 2008, system-wide roll-out was pushed back to 2011, with the MTA citing technical problems.[91] [92] However, in December 2009 an initial trial was implemented at three stations along the IRT Pelham Line (6 <6> trains) in the Bronx.[93] The announcements are voiced by radio traffic reporter Bernie Wagenblast[94] and Carolyn Hopkins, the voice of the New York subway.[95] As of March 9, 2011, the system is now active in about 110 stations all over the New York City Subway.[96] [97] [98] [99] [100] [101] [102]

RFID trial on the IRT Lexington Avenue Line

2006 PayPass only trial

The MTA signed a deal with MasterCard in the first few months of 2006 to test out a new RFID card payment scheme.[103] Customers had to sign up at a special MasterCard website and use a MasterCard PayPass credit or debit card/tag to participate. Participating stations included:[104]

- IRT Lexington Avenue Line (4 5 6 <6> trains) from both 138th Street stations in the Bronx to Borough Hall, Brooklyn,
- Court Square – 23rd Street in Queens for the E M trains,
- and Court Square for the 7 <7> trains.

Originally scheduled to end in December 2006, the MTA extended the trial due to "overwhelming positive response".[105]

2010 PayPass and PayWave trial

In light of the success of the first Paypass pilot project in 2006, another trial was started by the MTA. This one started on June 1, 2010 and ended on November 30, 2010. The first two months started with the customer just using the MasterCard PayPass debit or credit card.[106] [107] [108] [109] [110] However, this trial was the debut of having a rider use the VISA PayWave debit or credit card to enter the system, which started on August 1, 2010.[111] For six months, a rider could use either a MasterCard Paypass or VISA PayWave credit/debit card to pay for a fare on the following locations or routes:

- the IRT Lexington Avenue Line (4 5 6 <6> trains) from 138th Street – Grand Concourse or Third Avenue – 138th Street stations in the Bronx to Borough Hall station in Brooklyn,
- the M14, M23, M79, M86, M101, M102, M103 New York City Transit local bus routes, and the BxM7 MTA Bus express bus route,
- most of the Port Authority Trans-Hudson (PATH) train stations (except for Christopher Street and 9th Street); and,
- #6 (Ocean Avenue - Journal Square), #80 (Newark Avenue), and #87 (King Drive) New Jersey Transit bus routes.
- A new route service was added at an unknown date. Here, one can now try this trial to use the Newark Liberty International Airport's AirTrain monorail system to terminals A, B, and C and the long-term parking areas of the airport. However, it is only good for going to the airport, *away* from the Newark Liberty International Airport New Jersey Transit train station. This service does not apply heading to exit the airport and towards the mentioned Newark Liberty International Airport station.

Two options are available during this second trial for fare payment:

- "pay-as-you-go" RFID card scan at select turnstiles or locations; or,

- pre-funded fares via a pilot website called the "NY/NJ Transit Trial"[112] for multiple and unlimited ride discounts. Effective October 16, 2010, pre-funded fares are not available in the trial website.

Automation

In the late '90s and early 2000s, the MTA began the process of automating the subway. Beginning with the BMT Canarsie Line (L train) and the IRT Flushing Line (7 <7> trains),[113] the MTA has plans to eventually automate a much larger portion, using One Person Train Operation (OPTO) in conjunction with Communication-Based Train Control (CBTC). Siemens Transportation Systems built the CBTC system on the Canarsie line. Thales is building the CBTC system for the Flushing Line. In late winter 2008, the MTA embarked on a 5-week renovation and upgrade project on the 7 train between Flushing – Main Street and 61st Street – Woodside to upgrade signaling and tracks for CBTC. On February 27, 2008, the MTA issued an Accelerated Capital Program to continue funding the completion of CBTC for the 7 train and to begin on the IND Queens Boulevard Line (E F M R trains). The proposed plan is estimated to cost US $1.4 million.[114] (p. 15–16)

The New York City Subway uses a system known as Automatic Train Supervision (ATS) for dispatching and train routing on the A Division. ATS allows dispatchers in the Operations Control Center (OCC) to see where trains are in real time, and whether each individual train is running early or late. Dispatchers can hold trains for connections, re-route trains, or short-turn trains to provide better service when a disruption causes delays.

Help Point

The MTA set up another technology pilot project for the New York City Subway called "Help Point" in April 5, 2011. Help Point is a new digital-audio communications system that lets a rider access it, in case of an emergency or needing subway information for travel directions.[115] The top button is labeled red for emergencies towards the Rail Control Center. The bottom button is labeled green towards a MTA station agent for any inquiries. All units are equipped with a microphone to speak into and a speaker to hear answers to the rider from a MTA worker.[116] Also, the test units were equipped for the hearing impaired (under ADA compliance).

The two subway stations that were part of this trial were on the IRT Lexington Avenue Line. They were the 23rd Street and the Brooklyn Bridge – City Hall stations. The Help Points at the Brooklyn Bridge – City Hall station were wireless and the 23rd Street station ones were hard-wired, to test which type of transmission is best for the subway. If the test project is successful, then this communications system will replace the existing Customer Assistance Intercom (CAI) units on all 468 subway stations in the future.[117] [118]

On March 6, 2012, the MTA decided that all CAI units will be replaced with wireless Help Points in all subway stations, with optional cameras to each unit, where and when the money is available to have them installed.[119]

On The Go! Travel Station

The On the Go! Travel Station in use at the Bowling Green station

On September 19, 2011, the MTA set up another pilot project, an online, interactive touchscreen computer program called "On The Go! Travel Station" (OTG). It lists any planned work or service changes occurring on the subway as well as information to help travelers find landmarks or locales near the stations with an OTG outlet. The first station to test this new technology was Bowling Green on the IRT Lexington Avenue Line.[120] Other stations scheduled to participate in this program were Penn Station (with the LIRR), Grand Central Terminal (with Metro-North), Atlantic Avenue – Barclays Center in Brooklyn, and Jackson Heights – Roosevelt Avenue / 74th Street – Broadway in Queens.[121]

Safety and security

Crime, train accidents, suicides and threats of terrorism all impact the subway system.

Train movement safety

Safe train operation on the whole New York City Subway is ensured by a combination of interlocking, signalling, wayside train protection and wayside speed control layouts.[122] However, no technical system is free of hazards.

Interlocking

Like the railways, the subway used mechanical interlocking in early days and introduced relay interlocking later. Computer-based interlockings are state-of-the-art systems offering additional functions. But independently from the applied technology, the interlocking logic stayed the same: "Control lengths" along the selected route to be set until the target signal plus an additional overlap (safety distance) must be clear of any trains or cars to be able to clear the signal for the according route and target signal. "Single line signal diagrams" show all defined "control lengths" (and routes) for each interlocking tower.

Signalling

New York City Subway generally distinguishes signals into:

- automatic signals, controlled only by train movements
- approach signals, like automatic signals, can be forced to switch to stop aspect by interlocking tower
- home signals, route set by interlocking tower
- additional signals (call-on, dwarf, marker, sign, time signals)

Common automatic and approach signals consist of one signal head showing one of the following signal aspects:

- stop (one red light); with special rules for call-on and timer signals
- clear, next signal at clear or caution (one green light)
- proceed with caution, be prepared to stop at next signal (one yellow light)

Where different directions are possible, the subway uses both speed and route signalling, but with two different home signal aspect philosophies:

- IRT sections with older signalling:
  - one signal head for each possible route ahead (with main route above diverging routes)

- migrated IRT sections, BMT and IND divisions:
  - upper signal head for speeds
  - lower signal head for routes (with main route shown green and diverging route shown yellow)

Train protection

Subway trains are stopped mechanically at all signals showing "stop" aspects by automatic train stops located on the right side of IRT tracks and the left side of BMT/IND tracks; all cars are equipped with tripcocks. Although this is a simple principle of train stops, that wayside trippers must not be removed to trip ("stop") position until it is guaranteed that the train has fully passed the signal with all its cars.

Speed control

Speed control on the subway ensured by "Time Signals": A timer is started as soon as the train passes a certain point and will clear the signal ahead as soon as the predefined time elapsed; the minimum time is calculated from the speed limit and the distance between start of timer and signal. "Time Signals" are distinguished into "Grade Timer" for speed supervision at grades, curves or in front of buffer stops, and in "Station Timer" for low-speed entrance into stations in order to reduce train headways.

CBTC

Communication-Based Train Control (CBTC) is the state-of-the-art system for signalling, Automatic Train Protection and speed control being installed in the subway system within its current big automation project.

Train accidents

Including the predecessors of the New York City Subway, at least 55 train accidents have been recorded since 1918, when a train bound for South Ferry smashed into two trains halted near Jackson Avenue on the IRT White Plains Road Line in the Bronx.[123] The deadliest accident, the Malbone Street Wreck, happened November 1, 1918 beneath the intersection of Flatbush Avenue, Ocean Avenue, and Malbone Street near the Prospect Park station of the then-BRT Brighton Line in Brooklyn, killing 93 people.[124]

Passenger safety

Crime

The subway carries up to 1.5 billion passengers a year. Crime rates have shown variations over time, with a drop starting in the '90s and continuing today.[125] [126]

Yellow platform edges, yellow staircase steps and yellow railings, painted for safety

In order to fight crime, various approaches have been used.[127] A new initiative by the MTA to prevent crime is to ban people who commit one in the subway system from entering it for a certain length of time.[128] In the '60s, for example, Mayor Wagner ordered an increase in the Transit Police force from 1,219 to 3,100 officers. During the hours at which crimes most frequently occurred (between 8:00pm and 4:00 am), the officers went on patrol in all stations and trains. In response, crime rates decreased, as extensively reported by the press.[129]

In July 1985 however, the Citizens Crime Commission of New York City published a study showing riders abandoning the subway, fearing the frequent robberies and generally bad circumstances.[130]

To counter these developments, policy that was rooted in the late 1980s and early 1990s was implemented.[131] [132] In line with this *Fixing Broken Windows* philosophy, the New York City Transit Authority (NYCTA) began a five-year program to eradicate graffiti from subway trains in 1984.[133]

In 1989 the Metropolitan Transportation Authority asked the transit police (then located within the NYCTA) to focus on minor offenses such as fare evasion. In the early nineties, the NYCTA adopted similar policing methods for Penn Station and Grand Central Terminal.

In 1993, Mayor Rudy Giuliani took office and with Police Commissioner Howard Safir the strategy was more widely deployed in New York under the rubrics of "zero tolerance" and "quality of life". Crime rates in the subway and city dropped,[134] prompting *New York Magazine* to declare "The End of Crime as We Know It" on the cover of its August 14, 1995 edition.

Giuliani's campaign credited the success to the zero tolerance policy.[135] The extent to which his policies deserve the credit is disputed.[136]

New York City Police Department Commissioner William J. Bratton and author of *Fixing Broken Windows*, George L. Kelling, however, stated the police played an "important, even central, role" in the declining crime rates.[137] The trend continued and Giuliani's successor, Mayor Michael Bloomberg, stated in a November 2004 press release that "Today, the subway system is safer than it has been at any time since we started tabulating subway crime statistics nearly 40 years ago."[138]

Suicides

A portion of subway-related deaths in New York consists of suicides committed by jumping in front of an oncoming train. Exact numbers are not available, as the cause of death is listed as "unknown" in cases where no witness was present and no suicide note found.

Between 1990 and 2003, 343 subway-related suicides have been registered out of a citywide total of 7,394 (4.6%) and subway-related suicides increased by 30%, despite a decline in overall suicide numbers.[139]

Several planned stations in the New York City Subway may possibly feature platform screen doors. This includes the 7 line extension,[140] and Second Avenue Subway.[141] Although these doors are designed to improve airflow in stations, they also prevent people from falling or jumping onto the tracks.[142]

Photography

After the September 11 attacks in New York, the MTA was extremely wary of anyone taking photographs or recording video inside the system and proposed banning all photography and recording in a meeting around June 2004.[143] However, due to strong response from both the public and from civil rights groups, the rule of conduct was dropped. In November 2004, the MTA again put this rule up for approval, but was again denied,[144] though many police officers and transit workers still confront or harass people taking photographs or video.[145]

On April 3, 2009, the NYPD issued a directive to officers stating that it is legal to take pictures within the subway system so long as it is not accompanied with suspicious activity.[146]

Currently, the MTA Rules of Conduct,[59] Restricted Areas and Activities section states that anyone may take pictures or record video, provided that they do not violate MTA regulations:

> **Section 1050.9 Restricted areas and activities.**
>
> *Photography, filming or video recording in any facility or conveyance is permitted except that ancillary equipment such as lights, reflectors or tripods may not be used. Members of the press holding valid identification issued by the New York City Police Department are hereby authorized to use necessary ancillary equipment. All photographic activity must be conducted in accordance with the provisions of this Part.*[147]

Terrorism prevention and foiled plot

On July 22, 2005, in response to bombings in London, the New York City Transit Police introduced a new policy of randomly searching passengers' bags as they approached turnstiles. The NYPD claimed that no form of racial profiling would be conducted when these searches actually took place. The NYPD has come under fire from some groups that claim purely random searches without any form of threat assessment would be ineffectual. "This NYPD bag search policy is unprecedented, unlawful and ineffective," said Donna Lieberman, Executive Director of the NYCLU. "It is essential that police be aggressive in maintaining security in public transportation. But our very real concerns about terrorism do not justify the NYPD subjecting millions of innocent people to suspicionless searches in a way that does not identify any person seeking to engage in terrorist activity and is unlikely to have any meaningful deterrent effect on terrorist activity."[148] The searches were upheld by the United States Court of Appeals for the Second Circuit in *MacWade v. Kelly*.

On April 11, 2008, MTA received a Ferrara Fire Apparatus Hazardous Materials Response Truck, which went into service three days later. It will be used in the case of a chemical or bioterrorist attack.[149]

Najibullah Zazi and alleged co-conspirators were arrested in September 2009 and pled guilty to being part of an al-Qaeda plan to engage in suicide bombings on the New York City subway system during rush hour that month.[150] [150] [150] [151] [152] [153] [154] [155]

Challenges

2009–2010 budget cuts

The MTA faced a budget deficit of US$1.1 billion in 2009. Fare increases (three times from 2008 to 2010) and service reductions (including the elimination of two part-time subway services, the V and W, increased weekend headways on other services, and shortening, rerouting, or elimination of many bus routes) were enacted to balance the deficit.

Capacity constraints

The interior of an **F** train during morning rush hour

Several subway lines have reached their operational limits in terms of train frequency and passengers, according to data released by the Transit Authority. All but one of the A Division services as well as the E and L trains are beyond capacity, as well as portions of the N train.[156] [157]

The Second Avenue Subway will relieve pressure on the Lexington Avenue Line (4 5 6 <6>) when its first segment begins operating in 2016, but no such relief is planned for other crowded lines. However, the Long Island Railroad East Side Access project is expected to bring many more commuters to the Lexington Avenue Line at about the same time, further overwhelming its capacity. Because new subway construction can require years to plan and complete, the Transit Authority can only turn to increased bus service to manage demand in the short run, until automation of the subways using CBTC allows trains to run with less headway.

Subway flooding

Service on the subway system is occasionally disrupted by flooding from rainstorms, even minor ones.[158] Rainwater can disrupt signals underground and require the electrified third rail to be shut off. Since 1992, $357 million has been used to improve 269 pump rooms. As of August 2007, $115 million has been earmarked to upgrade the remaining 18 pump rooms.[159] Despite these improvements, the transit system continues to experience flooding problems.

On August 8, 2007, after more than 3 inches (mm) of rain fell within an hour, the subway system flooded, causing every line to either be disabled or seriously disrupted, effectively halting the morning rush. This was the third incident in 2007 in which rain disrupted service. The system was disrupted on this occasion because the pumps and drainage system can handle only a rainfall rate of 1.75 inches (mm) per hour; the incident's severity was aggravated by the scant warning as to the severity of the storm.[160] (p. 10) In late August 2007, MTA Engineer Phil Kollin announced new plans to create a system that would pump water away from the third rail.

On August 27, 2011, due to the approach of Hurricane Irene, the MTA suspended subway service at 12:00 noon in anticipation of heavy flooding on tracks and in tunnels. It was the first weather-caused shutdown in the history of the system.[161] Service was restored by Monday, August 29.[162] [163]

In addition, as part of a $130 million and an estimated 18 month project, the MTA began installing new subway grates in September 2008 in an attempt to prevent rain from overflowing into the subway system. The metallic structures, designed with the help of architectural firms and meant as a piece of public art, are placed atop existing grates but with a 3-to-4-inch (to mm) sleeve to prevent debris and rain from flooding the subway. The racks will at first be installed in the three most flood-prone areas as determined by hydrologists, including Jamaica, TriBeCa and the Upper West Side. Each neighborhood is scheduled to have its own distinct design, some featuring a wave-like deck which increases in height and features seating (Jamaica), others with a flatter deck that includes seating and a bike rack.[164] [165]

Hurricane Sandy service disruptions

On October 29, 2012 another full closure was ordered before the arrival of Hurricane Sandy. All services on the subway, the Long Island Rail Road and Metro-North were gradually shut down that day at 7:00 P.M., to protect passengers, employees and equipment from the coming storm. The storm caused serious damage to the system, especially the IND Rockaway Line, which had many sections between Howard Beach – JFK Airport and Hammels Wye on the Rockaway Peninsula heavily damaged, leaving it essentially isolated from the rest of the system. This required the NYCTA to truck in 20 subway cars to the line to provide some interim service (temporarily designated the H). Also, several of the system's tunnels under the East River were flooded by the storm surge.[166] South Ferry suffered serious water damage and is closed until further notice.[167]

Rain from drainage pipes comes into a subway car

Flooding at 125th Street after a water main break

Preparations for Hurricane Sandy at Bowling Green

South Ferry after Hurricane Sandy

Pump train in Cranberry Street Tunnel after Hurricane Sandy

Announcement about the temporary H shuttle after Hurricane Sandy

Litter and rodents

Heavily gratified subway car in New York City in 1973

Measures against rats

Litter accumulation is a perennial problem in the subway system. In the 1970s and 1980s, dirty trains and platforms, as well as graffiti were a serious problem. The situation has improved since then, but the 2010 budget crisis has threatened to curtail trash removal from the subway system.[168] [169]

The New York City Subway system is infested with rats.[170] [171] [172] Rats are sometimes seen on platforms,[173] and are commonly seen foraging through garbage thrown onto the tracks.[174] They are believed to pose a health hazard, and on rare instances have been known to bite humans.[175] Subway stations notorious for rat infestation include Chambers Street, Jay Street – MetroTech, West Fourth Street, and Spring Street.[176] Another station noted for rat infestations is the 5-car-only 145th Street station.[177]

Decades of efforts to eradicate or simply thin the rat population in the system have been unsuccessful. In March 2009, the Transit Authority announced a series of changes to its vermin control strategy, including new poison formulas and experimental trap designs.[178] In October 2011, the MTA announced a new initiative to clean 25 subway stations, along with their garbage rooms, of rat infestations.[179] Also in October 2011, the MTA announced a pilot program aimed at reducing levels of garbage in the subways by removing all garbage bins from the subway platforms. The initiative is being tested at the Eighth Street – New York University and Flushing – Main Street stations.[180]

Noise

New York subway trains produce high levels of noise that exceed guidelines set by the World Health Organization and U.S. Environmental Protection Agency.[181] In 2006, Columbia University's Mailman School of Public Health found noise levels averaged 95 decibel (dB) inside subway cars and 94 dB on platforms.[181] Daily exposure to noise at such levels for as little as 30 minutes can lead to hearing loss.[181] Noise on one in 10 platforms exceeded 100 dB.[181] Under WHO and EPA guidelines, noise exposure at that level is limited to 1.5 minutes.[181] A subsequent study by Columbia and the University of Washington found higher average noise level in the subway (80.4 dB) than commuter trains including the Port Authority Trans-Hudson (PATH) (79.4 dB), the Metro-North (75.1 dB) and Long Island Railroad (LIRR) (74.9 dB).[182] Small numerical increases on the decibel scale can translate into large increases in sound intensity; specifically, it is a logarithmic scale: sound at 95 dB is 10 times more intense than at 85 dB and 100 times more intense than at 75 dB, and so forth.[182] In the second study, peak subway noise registered at

102.1 dB.[182]

Public relations

The Board of Transportation and then New York City Transit Authority (MTA New York City Transit) has had numerous events that promote increased ridership of their transit system.

Miss Subways

From 1941 to 1976, the Board of Transportation/New York City Transit Authority sponsored the "Miss Subways" publicity campaign. In the musical *On the Town*, the character Miss Turnstiles is based on the Miss Subways campaign.[183] [184] In one scene, the musical shows three sailors taking an uptown train at Times Square.

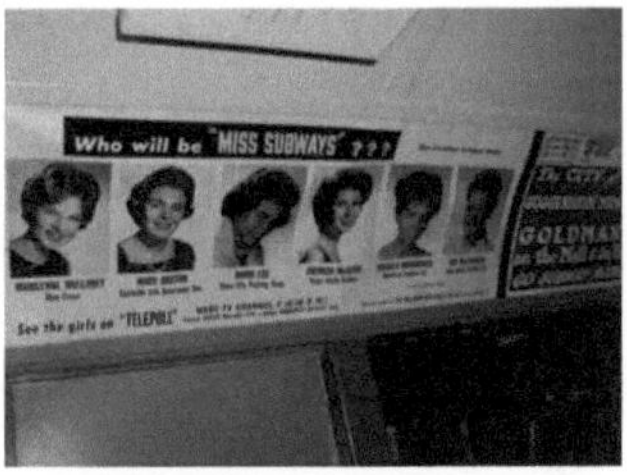

An advertisement for Miss Subways at the *New York City Transit Museum*

The campaign was resurrected in 2004, for one year, as "Ms. Subways". It was part of the 100th anniversary celebrations. Featuring young models, entertainers and others, the monthly campaign, which included the winners' photos and biographical blurbs on placards in subway cards, featured such winners as Mona Freeman and prominent New York City restaurateur Ellen Goodman. The winner of this contest was Caroline Sanchez-Bernat, an actress from Morningside Heights.[185]

Subway Series

Subway Series is a term attributed to any series of baseball games between New York City teams, called thus as opposing teams can travel to compete merely by using the subway system along with the fact that stations are adjacent and visible to their respective stadiums. Subway Series is a term long used in New York, going back to series between the Brooklyn Dodgers or New York Giants and the New York Yankees in the 1940s and '50s. Today, the term is used to describe the rivalry between the Yankees and the New York Mets. During the 2000 World Series, cars on the **4** train (which stopped at Yankee Stadium) were colored white with blue pinstripes, while cars on the **7** train (which stopped at Shea Stadium) were colored orange and blue, the Mets' team colors.

Holiday Special

Since 2003, the MTA has operated a Holiday Special on Sundays in November and December, from the first Saturday after Thanksgiving to the Saturday before Christmas Day.[186] This train was made of cars from the R1 through R9 series. The route made all stops between Second Avenue in Manhattan and Queens Plaza in Queens via the IND Sixth Avenue and IND Queens Boulevard Lines. In 2011, the train operated on Saturdays instead of Sundays.[187]

The contract, car numbers (and year built) used were: R1 100 (1930), R1 381 (1931), R4 401 (1932), R4 484 (1932) – Bulls Eye lighting and a test P.A. system added in 1946, R6-3 1000 (1935), R6-1 1300 (1937), R7A 1575 (1938) – rebuilt in 1947 as a prototype for the R10 subway car, R9 1802 (1940).

See also

- Graphical New York City Subway timeline
- List of metro systems
- List of United States rapid transit systems by ridership
- Metro systems by annual passenger rides
- Graffiti in New York
- Staten Island Railway – The rapid transit operator for Staten Island
- Subway Challenge
- New York City Subway in popular culture
- Transportation in New York City

References

[1] "Subway and Bus Ridership Statistics 2011" (http://mta.info/nyct/facts/ridership/index.htm). New York Metropolitan Transportation Authority. . Retrieved April 11, 2012.

[2] http://www.mta.info/nyct/

[3] The IRT main line, which is considered to be the first New York City "subway" line, opened in 1904; however, the Ninth Avenue Line, a predecessor elevated railroad line, operated its first trial run on July 3, 1868, according to *Facts and Figures 1979–80*, published by the New York City Transit Authority See also nycsubway.org (http://www.nycsubway.org/irt/9thave/), and the West End Line railroad opened in 1863. A small portion of the latter line's original right-of-way is still in daily use near Coney Island. thethirdrail.net (http://thethirdrail.net/9909/westend1.htm)

[4] "Facts & Figures" (http://www.nycsubway.org/faq/factsfigures.html). www.nycsubway.org. . Retrieved 2011-12-21.

[5] Hood, Clifton. 722 Miles. Baltimore, MD: Johns Hopkins University Press, 2004.

[6] "New York City Subway & Path" (http://www.urbanrail.net/am/nyrk/new-york.htm). www.urbanrail.net. . Retrieved November 7, 2010.

[7] New York State Comptroller. "A Guide for Evaluating the Metropolitan Transportation Authority's Proposed Capital Program for 2000 Through 2004" (http://www.osc.state.ny.us/osdc/rpt700/rpt700.htm). www.osc.state.ny.us. . Retrieved July 7, 2008.

[8] "MTA: Subway Ridership At Highest Level Since 1950" (http://www.ny1.com/content/news_beats/transit/159176/mta--subway-ridership-at-highest-level-since-1950). *NY1*. April 10, 2012. . Retrieved 2012-04-11.

[9] "(New York City) Subways (Facts and Figures under title of Annual Subway Ridership)" (http://www.mta.info/nyct/facts/ffsubway.htm). *MTA.info*. April 9, 2012. . Retrieved 2012-04-11.

[10] The Fulton Street and Broad Street stations of the BMT Nassau Street Line are closed on weekends. The Times Square and Grand Central stations of the IRT 42nd Street Shuttle are closed during late nights. The Aqueduct Racetrack station of the IND Rockaway Line is open only on live racing dates at Aqueduct Racetrack, in Ozone Park, Queens, from 11:00 AM to 7:00 PM.

[11] "About NYC Transit: Subways" (http://www.mta.info/nyct/facts/ffsubway.htm#stations). New York Metropolitan Transportation Authority. . Retrieved July 19, 2009.

[12] TRACK RECORD: 1.5B RODE SUBWAY" by Pete Donohue, New York Daily News, February 11, 2006

[13] Krauss, Clifford (May 10, 2008). "Gas Prices Send Surge of Riders to Mass Transit" (http://www.nytimes.com/2008/05/10/business/10transit.html). The New York Times. . Retrieved May 10, 2008.

[14] CBC: Subway ops efficient; buses, commuter rail not so much (http://secondavenuesagas.com/2011/04/07/cbc-subway-ops-efficient-buses-commuter-rail-not-so-much/)

[15] "Fact #67: BTU per Passenger Mile for U.S. Transit in 1995" (http://www1.eere.energy.gov/vehiclesandfuels/facts/favorites/fcvt_fotw67.html). *US Department of Energy. EERE Information Center*. www1.eere.energy.gov. September 4, 1998. . Retrieved July 7, 2008.

[16] "GHG Emissions, Urban Mobility and Efficiency of Urban Morphology" (http://www.urs2009.net/docs/papers/Bertaud.pdf). *Urban Research Symposium*. Marseille, France.: URS2009.net. May 14, 2009. . Retrieved November 24, 2009.

[17] http://www.youtube.com/watch?v=yZ83UhBJFP0

[18] http://www.youtube.com/watch?v=MsLCNw6HSMw

[19] "The Pneumatic Tunnel Under Broadway (1870)" (http://web.archive.org/web/20080505031528/http://www.nycsubway.org/articles/beach-1870-03-05.html). nycsubway.org. Archived from the original (http://www.nycsubway.org/articles/beach-1870-03-05.html) on May 5, 2008. . Retrieved July 7, 2008. (extract from *Scientific American* March 5, 1870)

[20] Mark S. Feinman. "History of the Independent Subway" (http://www.nycsubway.org/articles/historyindependentsubway.html). nycsubway.org. . Retrieved July 7, 2008.

[21] Kennedy, Randy. "Tunnel Vision; With Station's Reopening, Even Commuters Smile" (http://query.nytimes.com/gst/fullpage.html?res=9E02E1D81730F934A2575AC0A9649C8B63), *The New York Times*, September 17, 2002. Accessed October 6, 2007.

[22] Brian Abbott. September 11: Three Years Later. (http://brianabbott.net/observations/2004/09/11/september-11-three-years-later) Several maps showing changes in lower Manhattan from July 2001 to September 2002.

[23] "Types and Methods of Construction" (http://www.nycsubway.org/articles/irtbook_ch2.html). *IRT: The First Subway*. nycsubway.org. . Retrieved July 7, 2008.

[24] Subway Colors and Names (http://www.youtube.com/watch?v=yZ83UhBJFP0) MTA YouTube Web Page. Made July 15, 2010. Retrieved July 17, 2010.

[25] (http://mta.info/nyct/service/RebuildingRockawaysAfterHurricanSandy.htm), retrieved 12-3-2012.
[26] Polytechnic University Cable 34:2 Winter 2007 p.7
[27] Subway Service Changes (http://travel.mtanyct.info/serviceadvisory/default.aspx)
[28] New Subway Map is Here (http://mta.info/news/stories/?story=77) MTA.info website. Retrieved June 18, 2010.
[29] Updated New York City Subway Map (http://mta.info/maps/subway_web_full_map_Jun10.pdf) MTA.info website. Retrieved June 18, 2010.
[30] Late Night Subway Service (http://www.mta.info/maps/night_map_Sep12.pdf). Retrieved November 10, 2012.
[31] "The (Mostly) True Story of Helvetica and the New York City Subway" (http://www.aiga.org/content.cfm/the-mostly-true-story-of-helvetica-and-the-new-york-city-subway?pp=1). *AIGA*. aiga.org. November 18, 2008. . Retrieved February 4, 2009.
[32] Hogarty, Dave (August 3, 2007). "Michael Hertz, Designer of the NYC Subway Map" (http://gothamist.com/2007/08/03/michael_hertz_d.php). *Gothamist*. . Retrieved July 4, 2009.
[33] "Introducing The Weekender" (http://www.youtube.com/watch?v=c30CwkKyTYQ). *MTA.info (YouTube)*. September 30, 2011. . Retrieved 2011-10-01.
[34] "The Weekender" (http://mta.info/index.html). *MTA.info*. September 16, 2011. . Retrieved 2011-09-18.
[35] Grynbaum, Michael M. (September 15, 2011). "Aid for Baffled Weekend Subway Riders" (http://www.nytimes.com/2011/09/16/nyregion/new-subway-map-to-help-riders-with-weekend-service-changes.html?_r=1&ref=nyregion). *The New York Times*. . Retrieved 2011-09-30.
[36] "Introducing 'The Weekender'" (http://mta.info/news/stories/?story=384). *MTA.info*. September 16, 2011. . Retrieved 2011-09-18.
[37] "MTA Launches Interactive Online Map Ahead Of Difficult Weekend For Subways" (http://www.ny1.com/content/news_beats/transit/147316/mta-launches-interactive-online-map-ahead-of-difficult-weekend-for-subways). *NY1*. September 16, 2011. . Retrieved 2011-09-18.
[38] "MTA Releases "Weekender" Smartphone App" (http://www.ny1.com/content/news_beats/transit/162875/mta-releases--weekender--smartphone-app). *NY1*. June 11, 2012. . Retrieved 2012-06-13.
[39] "'The Weekender' Is Now an iPhone App" (http://mta.info/news/stories/?story=707). *MTA.info*. June 11, 2012. . Retrieved 2012-06-13.
[40] Subway Map Gets A Makeover (http://www.ny1.com/content/top_stories/119467/subway-map-gets-a-makeover) NY1 local news channel. Retrieved May 28, 2010.
[41] TRNSPRTNATION | New York City (http://www.trnsprtnation.com/nyc.html). Retrieved August 26, 2011.
[42] Artist Turns New York Subway Map Into a Musical Instrument (http://www.good.is/post/new-york-subway-map-turned-into-a-musical-instrument/). Retrieved August 26, 2011.
[43] Kennedy, Randy (August 13, 2002). "Tunnel Vision; The Light at the Top of the Subway Stairs" (http://query.nytimes.com/gst/fullpage.html?res=980DE2DF113AF930A2575BC0A9649C8B63). *The New York Times*. . Retrieved July 7, 2008.
[44] "Arts for Transit: The Official Subway and Rail Art Guide" (http://www.mta.info/mta/aft/). *Metropolitan Transportation Authority*. www.mta.info. . Retrieved July 7, 2008.
[45] Subway Art Hard To Miss, Easy To Uncover (http://www.ny1.com/content/news_beats/transit/128782/subway-art-hard-to-miss--easy-to-uncover/) NY1 local news channel. Made November 11, 2010. Retrieved November 12, 2010.
[46] Subway Art Guide (http://www.nycsubway.org/perl/artwork)
[47] Smith, Roberta (January 2, 2004). "The Rush-Hour Revelations of an Underground Museum" (http://www.nytimes.com/2004/01/02/arts/design/02XSMIT.html?ex=1073797200&en=453730e9e43c8edd&ei=5040&partner=MOREOVER). *The New York Times* (New York, NY).
[48] "MTA Guide to Accessible Transit" (http://www.mta.info/accessibility/index.html). New York Metropolitan Transportation Authority. . Retrieved March 8, 2010.
[49] "MTA Guide to Accessible Transit: MTA Subways and Buses" (http://www.mta.info/accessibility/transit.htm). New York Metropolitan Transportation Authority. . Retrieved March 8, 2010.
[50] "About New York City Transit" (http://mta.info/nyct/facts/about_us.htm). *MTA.info*. . Retrieved 2011-06-01.
[51] "MTA Guide to Accessible Transit" (http://mta.info/accessibility/stations.htm). *MTA.info*. . Retrieved 2011-06-01.
[52] MTA's Arts for Transit's Music Under New York (MUNY) Section (http://www.mta.info/mta/aft/muny/) MTA.info website.
[53] "2010 Music Under New York Auditions" (http://www.youtube.com/watch?v=9K5HpspsQcA). *MTA.info (YouTube)*. May 27, 2010. . Retrieved 2010-05-29.
[54] Redwine, Tina (May 18, 2011). "MTA Holds Tryouts For Subway Performers" (http://www.ny1.com/content/139318/mta-holds-tryouts-for-subway-performers). *NY1*. . Retrieved 2011-05-20.
[55] "2011 Music Under New York Auditions" (http://www.youtube.com/watch?v=zHbAGwoUIv0). *MTA.info (YouTube)*. May 27, 2011. . Retrieved 2011-05-28.
[56] "Performers Audition For Coveted MTA Subway Slots" (http://www.ny1.com/content/news_beats/transit/161317/performers-audition-for-coveted-mta-subway-slots). *NY1*. May 16, 2012. . Retrieved 2012-05-16.
[57] "2012 Music Under New York Auditions" (http://www.youtube.com/watch?v=zDHDbfddk44). *MTA.info (YouTube)*. May 16, 2012. . Retrieved 2012-05-16.
[58] "M. Salieu Suso, Kora Player" (http://www.youtube.com/watch?v=0mzb31JkIi8). *MTA.info (YouTube)*. April 8, 2010. . Retrieved 2011-04-09.
[59] MTA Rules of Conduct (http://mta.info/nyct/rules/rules.htm)

[60] Haddon, Heather; Klopsis, Nicholas (April 11, 2010). "Flush with filth: Many subway station bathrooms dirty, or locked up" (http://amny.com/urbanite-1.812039/flush-with-filth-many-subway-station-bathrooms-dirty-or-locked-up-1.1857262). *AM New York*. . Retrieved April 16, 2010.

[61] "Rapid Transit Challenge-restroom locations" (http://www.rapidtransitchallenge.com/finish.htm). . Retrieved April 13, 2007.

[62] Trendy Shops in the Subway (http://mta.info/news/stories/?story=93) MTA.info website. Made August 26, 2010. Retrieved September 2, 2010.

[63] "New York City Subway Car Fleet June 2010 through June 2011" (http://www.thejoekorner.com/cars/cars-current.htm). TheJoeCorner. . Retrieved November 12, 2011.

[64] "R160A in service on the N on January 19, 2007" (http://www.nycsubway.org/perl/show?59931). .

[65] "R160B in service on the N on February 23, 2007" (http://www.nycsubway.org/perl/show?59998). .

[66] *AM New York* (Sept. 25, 2006): "Voices Down Below", by Justin Rocket Silverman

[67] "MTA On Track To Purchase New Subway Cars" (http://www.ny1.com/content/news_beats/transit/158331/mta-on-track-to-purchase-new-subway-cars). *NY1*. March 27, 2012. . Retrieved 2012-03-27.

[68] "Governor Cuomo Announces $600 Million MTA Investment in Upstate Manufacturing (MTA Chooses Bombardier for $600 Million Order for 300 Subway Cars, More Than 500 Manufacturing and Supply Chain Jobs Maintained or Created)" (http://www.governor.ny.gov/press/032812MTA-Investment). *Government of New York*. March 28, 2012. . Retrieved 2012-03-29.

[69] Cudahy, Brian (2003). *A Century of Subways: Celebrating 100 Years of New York's Underground Railways*. Fordham University Press. ISBN 0-8232-2292-6., p. 28

[70] New York City Subway Tokens (http://www.nycsubway.org/wiki/New_York_City_Subway_Tokens)

[71] "TUNNEL VISION; The Kiss of Desperation: A Disgusting Practice Vanishes With the Token" (http://www.nytimes.com/2003/04/08/nyregion/08TUNN.html?scp=3&sq=TUNNEL VISION; The Kiss of Desperation: A Disgusting Practice Vanishes With the Token&st=cse) by Randy Kennedy, *The New York Times*, April 8, 2003

[72] Heilbroner, David (1991). "ECAB". *Rough justice*. New York: Dell. p. 14. ISBN 0-440-21030-5.

[73] "Connecticut Highway token buys subway ride, too", *The New York Times*, November 18, 1982, p.1

[74] "Connecticut to alter its turnpike tokens, solving subway issue", *The New York Times*, December 15, 1982, p.1

[75] "17½ Accord puts an end to the great token war", *The New York Times*, November 7, 1985,

[76] Newman, Andy (July 3, 1998). "Hop On, Hop Off: The Unlimited Metrocard Arrives" (http://www.nytimes.com/1998/07/03/nyregion/hop-on-hop-off-the-unlimited-metrocard-arrives.html). *The New York Times*. . Retrieved January 8, 2010.

[77] "Capital Construction Major Projects" New York City Transit (http://www.mta.info/capconstr/) (accessed Jan. 13, 2009)

[78] MTA Board Approves Capital Construction Plan (http://www.ny1.com/5-manhattan-news-content/top_stories/117723/mta-board-approves-capital-construction-plan) NY1 local news channel. Retrieved April 28, 2010.

[79] MTA Capital Program 2010–2014 (http://mta.info/news/stories/?story=52) MTA.info website. Retrieved June 25, 2010.

[80] MTA Board Reconsiders Billions Spent On Capital Budget (http://www.ny1.com/content/122992/mta-board-reconsiders-billions-spent-on-capital-budget) NY1 local news channel. Made July 30, 2010. Retrieved July 31, 2010.

[81] Tracking the Capital Program (http://mta.info/news/stories/?story=90) MTA.info website. Retrieved August 11, 2010.

[82] Capital Program Dashboard 2010 - 2014 Plan (http://www.mta.info/capitaldashboard/10_14/CapitalDashBoard7.html) MTA.info website. Retrieved August 11, 2010.

[83] "MTA to merge agencies into five companies" (http://www.progressiverailroading.com/news/article.asp?id=10332). October 11, 2002. . Retrieved 2009-03-07.

[84] "Completed Portions of the 2nd Ave. Subway" (http://www.nycsubway.org/articles/2ndave-builtfaq.html). nycsubway.org. . Retrieved July 7, 2008.

[85] "MTA FASTRACK page" (http://mta.info/nyct/service/fastrack_changes.htm). MTA. . Retrieved May 13, 2012.

[86] "MTA Youtube video about FASTRACK" (http://www.youtube.com/watch?v=g67pFDTZyZs). MTA. . Retrieved May 13, 2012.

[87] "MTA New York City Transit Begins Using Video Screens in Pilot Program to Provide Real-Time Information on the L Line" (http://www.mta.info/mta/news/releases/?en=090205-NYCT18). MTA. February 5, 2009. . Retrieved February 9, 2009.

[88] "Delay On Subway Message Boards, City Looks For New Contractors" (http://www.ny1.com/ny1/content/index.jsp?stid=5&aid=63640). *NY1 news*. www.ny1.com. October 20, 2006. . Retrieved July 7, 2008.

[89] Olshan, Jeremy (January 26, 2007). "ABOUT TIME FOR MTA" (http://www.nypost.com/seven/01262007/news/regionalnews/about_time_for_mta_regionalnews_jeremy_olshan.htm). *New York Post*. . Retrieved July 7, 2008.

[90] "No ETA: Firm flubs $160M Subway Tech" (http://www.nypost.com/seven/10202006/news/regionalnews/no_eta__firm_flubs_160m_subway_tech_regionalnews_jeremy_olshan__transit_reporter.htm). *New York Post* (New York Post). October 20, 2006. . Retrieved July 7, 2008.

[91] Pete Donohue (September 23, 2008). "Subway arrival info behind schedule" (http://www.nydailynews.com/ny_local/2008/09/23/2008-09-23_subway_arrival_info_behind_schedule.html). *New York Daily News*. . Retrieved September 24, 2008.

[92] "MTA Communications Network Unfinished, Out Of Date" (http://www.ny1.com/content/news_beats/transit/134850/mta-communications-network-unfinished--out-of-date/). *NY1*. March 2, 2011. . Retrieved 2011-03-02.

[93] "Subway Countdown Clocks Pass the Test" (http://wcbstv.com/video/?id=136436@wcbs.dayport.com). *CBS2 New*. December 24, 2009. . Retrieved December 31, 2009.

[94] "Heard Underground, in Your Dreams" (http://cityroom.blogs.nytimes.com/2012/02/23/your-fantasy-subway-announcements/?ref=nyregion). *New York Times*. February 23, 2012. . Retrieved 2011-02-28.

[95] Announcing the Subway Announcement Lady (http://cityroom.blogs.nytimes.com/2010/11/14/announcing-the-subway-announcement-lady/?hp). // The New York Times, 14.11.2010

[96] "Countdown Clocks" Continue to Pop Up on Subway Platforms (http://mta.info/news/stories/?story=76) MTA.info website. Retrieved June 18, 2010.

[97] Blocked Subway Signs A Tough Read, Straphangers Say (http://www.ny1.com/content/124195/blocked-subway-signs-a-tough-read--straphangers-say) NY1 local news channel. Retrieved August 23, 2010.

[98] Not All Riders Feel The Time Is Right For Subway Station Countdown Clocks (http://www.ny1.com/content/124546/not-all-riders-feel-the-time-is-right-for-subway-station-countdown-clocks) NY1 local news channel. Retrieved August 28, 2010.

[99] Subway Countdown Clock Pilot (http://www.youtube.com/watch?v=_a_CA5wiFgU) MTA's YouTube website. Made February 26, 2010. Retrieved September 1, 2010.

[100] Train Arrival Information Signs (http://www.youtube.com/watch?v=mTWxH-I3aUY) MTA's YouTube website. Retrieved October 9, 2010.

[101] The Clock Strikes 100! (http://www.mta.info/news/stories/?story=164) MTA.info website. Made December 21, 2010. Retrieved December 23, 2010.

[102] "Countdown Clocks Go Express.....or Local" (http://www.mta.info/news/stories/?story=199). MTA.info. March 9, 2011. . Retrieved 2011-03-10.

[103] "NYC Subway-Mastercard Trial" (http://www.engadget.com/2006/02/01/select-customers-to-trial-rfid-nyc-subway-pass/). engadget.com. February 1, 2006. . Retrieved April 9, 2007.

[104] "Mastercard Paypass NYC Subway Trial-Stations" (http://web.archive.org/web/20070528230220/http://www.mastercard.com/us/paypass/subway/about/wheretotap.html). Mastercard. Archived from the original (http://www.mastercard.com/us/paypass/subway/about/wheretotap.html) on May 28, 2007. . Retrieved April 9, 2007.

[105] "Extension of Paypass Trial (dead link)" (http://web.archive.org/web/20070502191816/http://www.mastercard.com/us/paypass/subway/faresoverview/autopayplan.html). Mastercard. Archived from the original (http://www.mastercard.com/us/paypass/subway/faresoverview/autopayplan.html) on May 2, 2007. . Retrieved April 9, 2007.

[106] MTA Launches Smart Card Pilot Program (http://mta.info/news/stories/?story=70) - MTA.info website. Retrieved June 2, 2010.

[107] Pilot Program Replaces MetroCard Swipe With Credit Card Tap (http://www.ny1.com/content/119590/pilot-program-replaces-metrocard-swipe-with-credit-card-tap) NY1. Retrieved June 2, 2010.

[108] Contactless Fare Payment Pilot (http://www.youtube.com/watch?v=UbhvC-4AEUE) - MTA's YouTube website. Retrieved June 1, 2010.

[109] MasterCard Tap & Go Payment System Enhances Commuter Experience (http://www.youtube.com/watch?v=RY4axTP_XyE) - Mastercard Worldwide YouTube website. Made May 28, 2010. Retrieved June 3, 2010.

[110] City Critic: A Quick Kiss of Plastic, and Ready to Ride (http://www.nytimes.com/2010/06/13/nyregion/13critic.html) New York Times. Made June 11, 2010. Retrieved June 14, 2010.

[111] NY/NJ Transit Trial FAQ #5 (http://www.ridenewyorknewjersey.com/faqs.html#a5) Retrieved October 16, 2010.

[112] NY/NJ Transit Trial Website (http://www.ridenewyorknewjersey.com/intro.html) Mastercard. Retrieved June 1, 2010.

[113] Arden, Patrick; Ventura, Michael P. (04-12). "Ready to roll on Second Ave: MTA's chief engineer dishes the dirt on their new dig" (http://ny.metro.us/metro/local/article/Ready_to_roll_on_Second_Ave/7934.html) (– Scholar search (http://scholar.google.co.uk/scholar?hl=en&lr=&q=intitle:Ready+to+roll+on+Second+Ave+<small>M T A â s+chief+engineer+dishes+the+dirt+on+their+new+dig</small>&as_publication=&as_ylo=2007&as_yhi=2007&btnG=Search)). *NY Metro*: pp. 1/2.

[114] "2008–2013 MTA Capital Program Accelerated Program Presentation to the Board" (http://mta.info/mta/news/public/ppt/Accelerated Program Presentation for Boardjss2.ppt) (PPT). Metropolitan Transportation Authority (New York). February 27, 2008. . Retrieved February 28, 2008.

[115] "Help Point Intercom Pilot" (http://www.youtube.com/watch?v=QTormQCbEcM). *MTA.info (YouTube)*. September 27, 2010. . Retrieved 2011-04-07.

[116] "Help Points Debut at Two Subway Stations" (http://mta.info/news/stories/?story=221). *MTA.info*. April 5, 2011. . Retrieved 2011-04-06.

[117] "MTA Kicks Off Test Program With New Subway Station Communication Units" (http://www.ny1.com/content/136875/mta-kicks-off-test-program-with-new-subway-station-communication-units). *NY1*. April 5, 2011. . Retrieved 2011-04-06.

[118] "Introducing the Help Point" (http://www.youtube.com/watch?v=j8k4mC88dQE). *MTA.info (YouTube)*. April 6, 2011. . Retrieved 2011-04-06.

[119] Redwine, Tina (March 6, 2012). "MTA To Provide Every Subway Station With "Help Point" Intercom" (http://www.ny1.com/content/news_beats/transit/157159/mta-to-provide-every-subway-station-with--help-point--intercom). *NY1*. . Retrieved 2012-03-08.

[120] "MTA's First "On The Go" Touchscreen Unveiled In Bowling Green" (http://www.ny1.com/content/top_stories/147462/mta-s-first--on-the-go--touchscreen-unveiled-in-bowling-green). *NY1*. September 19, 2011. . Retrieved 2011-10-05.

[121] "MTA Introduces New On the Go! Touch-Screen Travel Station" (http://mta.info/news/stories/?story=389). *MTA.info*. September 19, 2011. . Retrieved 2011-10-05.

[122] "NYC Subway Signals" (http://nycsubway.org/articles/signals.html). nycsubway.org. . Retrieved November 19, 2011.

[123] "NYC Subway accidents" (http://www.nycsubway.org/faq/accidents.html). nycsubway.org. 2009. . Retrieved December 12, 2009.
[124] "The Mayor to Begin B.R.T Inquiry Today" (http://query.nytimes.com/mem/archive-free/pdf?_r=1&res=9804E5DE1239E13ABC4C53DFB7678383609EDE) (PDF). *The New York Times* (New York City, NY): pp. 24. November 1, 1918. . Retrieved December 12, 2009.
[125] Redwine, Tina (November 14, 2011). "Transit Police: Subway Crime More Likely During Daytime, Manhattan Has Most Incidents" (http://www.ny1.com/content/news_beats/transit/150780/transit-police--subway-crime-more-likely-during-daytime--manhattan-has-most-incidents). *NY1*. . Retrieved 2011-11-15.
[126] Redwine, Tina (January 23, 2012). "More Transit Police Officers Combat Increase In Subway Crime" (http://www.ny1.com/content/news_beats/transit/154724/more-transit-police-officers-combat-increase-in-subway-crime). *NY1*. . Retrieved 2012-01-24.
[127] "If You See Something, Say Something - Subway - Spring 2011" (http://www.youtube.com/watch?v=z8eoTCEn2bY). *MTA.info (YouTube)*. April 19, 2011. . Retrieved 2011-04-19.
[128] Redwine, Tina (March 26, 2012). "MTA Proposes To Ban Those Who Commit Subway Crimes From Riding System" (http://www.ny1.com/content/news_beats/transit/158315/mta-proposes-to-ban-those-who-commit-subway-crimes-from-riding-system). *NY1*. . Retrieved 2012-03-27.
[129] Chaiken, Jan M.; Lawless, M.; Stevenson, K.A. (1974). "The Impact of Police Acitivity on Crime" (http://www.rc.rand.org/pubs/reports/R1424/). *RAND Corporation*.
[130] "Downtown Safety Security and Economic Development (Paperback)" (http://openlibrary.org/b/OL2741748M/Downtown_safety_security_and_economic_development). *Downtown Safety Security and Economic Development*. New York, NY, USA: Downtown Research & Development. 1986. pp. 148. . Retrieved November 24, 2009.
[131] Kelling, George L. (2009). "How New York Became Safe: The Full Story" (http://www.city-journal.org/2009/nytom_ny-crime-decline.html). City Journal. . Retrieved November 24, 2009.
[132] Glazer, Nathan (1979). "On Subway Graffiti in New York" (http://www.nationalaffairs.com/doclib/20080528_197905401onsubwaygraffitiinnewyorknathanglazer.pdf). *National Affairs* (54): pp. 3–12. . Retrieved November 24, 2009
[133] "DON'T DO IT (1988)" (http://www.youtube.com/watch?v=L4gq_wnEsmI). *MTA.info (YouTube)*. January 9, 2010. . Retrieved 2011-05-28.
[134] Wolff, Craig (November 12, 1993). "Subway Crime Declining, New Transit Figures Show" (http://www.nytimes.com/1993/11/12/nyregion/subway-crime-declining-new-transit-figures-show.html). *New York Times* (New York, NY, USA). . Retrieved November 26, 2009.
[135] "Rudy Giuliani" (http://web.archive.org/web/20070904220218/http://www.joinrudy2008.com/issues/). JoinRudy2008.com. Archived from the original (http://www.joinrudy2008.com/issues/) on September 4, 2007. . Retrieved July 15, 2007.
[136] Levitt, Steven D.: "Understanding Why Crime Fell in the 1990s: Four Factors that Explain the Decline and Six that Do Not", *Journal of Economic Perspectives*, 18(1), 163–190
[137] Kelling, George L.; Bratton, William J. (1998). "Declining crime rates: insiders' view of the New York City story" (http://www.jstor.org/pss/1144255). *The Journal of Criminal Law and Criminology (1973-)* (Northwestern University) **88** (4): pp. 1217–1220. . Retrieved November 26, 2009
[138] "Mayor Michael R. Bloomberg testifies before the Metropolitan Transportation Authority Board" (http://www.nyc.gov/portal/site/nycgov/menuitem.c0935b9a57bb4ef3daf2f1c701c789a0/index.jsp?pageID=mayor_press_release&catID=1194&doc_name=http://www.nyc.gov/html/om/html/2004b/pr298-04.html&cc=unused1978&rc=1194&ndi=1) (Press release). November 9, 2004. . Retrieved November 26, 2009.
[139] Galea, Sandro; Tracy, M; Piper, T.M.; Bucciarelli, A.M.; Tardiff, K.; Gershon, R; Vlahov, D (November 4, 2009). "Epidemiology of suicide in the New York City subway system" (http://apha.confex.com/apha/134am/techprogram/paper_133195.htm). American Public Health Association. . Retrieved November 27, 2009.
[140] MTA Plans To Install Platform Doors On 7 Line Extension (http://www.ny1.com/ny1/content/index.jsp?stid=1&aid=73389), NY1, September 8, 2007
[141] Glass walls, sliding doors on 2nd Ave. subway? (http://abclocal.go.com/wabc/story?section=local&id=5185580), WABC-TV, April 5, 2007
[142] "Mind the Gap" (http://www.railway-technology.com/features/feature49489/). February 10, 2009.
[143] Matt Haber (June 1, 2004). "Forbidden Photos, Anyone?" (http://www.villagevoice.com/nyclife/0423,haber,54075,15.html). www.villagevoice.com. . Retrieved July 7, 2008.
[144] "Threat Of Subway Photo Ban Riseth Again" (http://gothamist.com/2004/11/30/threat_of_subway_photo_ban_riseth_again.php). *Gothamist*. November 30, 2004. . Retrieved July 7, 2008.
[145] "Some Police Still Enforcing Non-Existent Ban On Photography In The Subways" (http://www.ny1.com/ny1/content/index.jsp?stid=5&aid=57334). *NY1 news*. www.ny1.com. February 23, 2006. . Retrieved July 7, 2008.
[146] 4-3-09 Directive 14: Railfans, use this for your own defense when confronted. (http://farm3.static.flickr.com/2337/3544963391_7f1d30700c_b.jpg)
[147] Full section (http://mta.info/nyct/rules/rules.htm#restricted)
[148] "NYCLU Sues New York City Over Subway Bag Search Policy" (http://www.aclu.org/police/searchseizure/20054prs20050804.html) (Press release). American Civil Liberties Union. August 4, 2005. .
[149] "MTA NYC Transit Introduces New Hazmat Response Vehicle" (http://www.railpace.com/hotnews/). April 11, 2008. . Retrieved April 14, 2008.

[150] Karen Zraik, David Johnston (September 15, 2009). 21, 2009 "Man in Queens Raids Denies Any Terrorist Link" (http://www.webcitation.org/query?url=http://www.nytimes.com/2009/09/16/nyregion/16terror.html&date=September). *The New York Times*. Archived from the original (http://www.nytimes.com/2009/09/16/nyregion/16terror.html?_r=1) on September 21, 2009. 21, 2009.

[151] Zazi Reveals Details Of Foiled Terror Plot (http://www.ny1.com/1-all-boroughs-news-content/top_stories/116757/zazi-reveals-details-of-foiled-terror-plot) NY1 local news channel. Retrieved April 12, 2010.

[152] Zazi Accomplice Pleads Guilty To Terror Charges (http://www.ny1.com/1-all-boroughs-news-content/top_stories/117456/zazi-accomplice-pleads-guilty-to-terror-charges) NY1 local news channel. Retrieved April 23, 2010.

[153] Report: NYPD Criticizes Port Authority Police's Failure To Stop Zazi (http://www.ny1.com/5-manhattan-news-content/top_stories/117576/report--nypd-criticizes-port-authority-police-s-failure-to-stop-zazi) NY1 local news channel. Retrieved April 26, 2010.

[154] Officials: Al-Qaida Terrorist Met With Subway Bomb Plotter (http://www.ny1.com/content/top_stories/121379/officials--al-qaida-terrorist-met-with-subway-bomb-plotter) NY1 local news channel. Retrieved July 1, 2010.

[155] Top Al-Qaida Leader Charged In Subway Bomb Plot (http://www.ny1.com/content/top_stories/121689/top-al-qaida-leader-charged-in-subway-bomb-plot) NY1 local news channel. Retrieved July 7, 2010.

[156] Neuman, William (June 26, 2007). "Some Subways Found Packed Past Capacity" (http://www.nytimes.com/2007/06/26/nyregion/26mta.html). *The New York Times*. . Retrieved July 7, 2008.

[157] Appendix A: Service Needs Assessment Methodology (http://www.nyc.gov/html/brt/downloads/pdf/brt_phase2_appendix.pdf).

[158] Latest Rainstorm No Match For City's Subway Pumps (http://www.ny1.com/content/news_beats/transit/126428/latest-rainstorm-no-match-for-city-s-subway-pumps/) NY1 local news channel. Retrieved October 2, 2010.

[159] Donohue, Pete (August 9, 2007). "Downpour swamps subways, stranding thousands of riders" (http://www.nydailynews.com/news/2007/08/09/2007-08-09_its_transit_hell_from_heavens.html). *New York Daily News*. . Retrieved August 23, 2007

[160] "August 8, 2007 Storm Report" (http://mta.info/mta/pdf/storm_report_2007.pdf) (PDF). Metropolitan Transportation Authority (New York). September 20, 2007. . Retrieved October 27, 2007.

[161] Aaron Feis, Sabrina Ford and Jennifer Fermino (August 27, 2011). "Hurricane Irene halts NY, NJ mass transit" (http://www.nypost.com/p/news/local/mass_transit_shutdown_mode_ge2t81TM8fgggFhhBkbHoJ). *The New York Post*. . Retrieved August 28, 2011.

[162] "NYC's subway service running OK Monday after Irene" (http://www.washingtonpost.com/lifestyle/travel/nycs-subway-service-running-ok-monday-after-irene/2011/08/29/gIQA1NnInJ_story.html?wpisrc=nl_headlines). *The Washington Post*. August 29, 2011. . Retrieved August 29, 2011.

[163] Clark, Roger (August 29, 2011). "Straphangers Have Relatively Easy Monday Morning Rush" (http://www.ny1.com/content/top_stories/145962/straphangers-have-relatively-easy-monday-morning-rush). *NY1*. . Retrieved 2011-08-30.

[164] Dunlap, David W. (September 19, 2008). "New Subway Grates Add Aesthetics to Flood Protection" (http://cityroom.blogs.nytimes.com/2008/09/19/new-subway-grates-add-aesthetics-to-flood-protection/). *The New York Times*. . Retrieved October 7, 2008.

[165] Lee, Jennifer (October 1, 2008). "Three in One — Flood Protection, Benches and Bike Parking" (http://cityroom.blogs.nytimes.com/2008/10/01/three-in-one-flood-protection-benches-and-bike-parking-in-a-new-design/). The New York Times. . Retrieved October 7, 2008.

[166] http://www.youtube.com/watch?v=mhJrrGNvcFk

[167] Restoring South Ferry Station (http://www.mta.info/nyct/service/RestoringSouthFerryStation.htm)

[168] MTA Staff Cuts May Boost Transit Trash (http://www.ny1.com/1-all-boroughs-news-content/top_stories/118296/mta-staff-cuts-may-boost-transit-trash) - NY1 local news channel. Retrieved May 7, 2010.

[169] "MTA Falls Behind On Trash Collection" (http://www.ny1.com/content/news_beats/transit/141746/mta-falls-behind-on-trash-collection). *NY1*. June 27, 2011. . Retrieved 2011-06-27.

[170] "NYC takes aim at subway rats" (http://www.upi.com/Top_News/2009/03/24/NYC-takes-aim-at-subway-rats/UPI-79751237928283/). *United Press International*. March 24, 2009. . Retrieved June 27, 2009.

[171] Transit Officials To Tackle Growing Subway Rat Problem (http://www.ny1.com/content/120413/subway-rat-problem-dominates-city-health-dept--meeting) - NY1 local news channel. Retrieved June 16, 2010.

[172] Farinacci, Amanda (January 5, 2012). "Transit Workers Call For Action Against Rats In Subway Stations" (http://www.ny1.com/content/news_beats/transit/153718/transit-workers-call-for-action-against-rats-in-subway-stations). *NY1*. . Retrieved 2012-01-08.

[173] " Rats rule the subway rails, platforms" (http://www.newsday.com/long-island/transportation/rats-rule-the-subway-rails-platforms-1.883713), Matthew Sweeney, *Newsday*, August 13, 2008

[174] Straphangers' Habits Feeding Subway Rat Population, Officials Say (http://www.ny1.com/content/top_stories/129581/straphangers--habits-feeding-subway-rat-population--officials-say) NY1 local news channel. Made November 25, 2010. Retrieved November 27, 2010.

[175] Beja, Marc (September 8, 2011). "Rat bites woman in subway station as she waits for train" (http://www.amny.com/urbanite-1.812039/rat-bites-woman-in-subway-station-as-she-waits-for-train-1.3152338). *amNewYork*. . Retrieved 2011-09-12.

[176] " Rats in subway are health hazard for New Yorkers" (http://www.brooklyntoday.info/component/content/article/2-general-news/152-rats-in-subway-are-health-hazard-for-new-yorkers.html), Pamela Branch, *Brooklyn Today*, January 12, 2010

[177] Redwine, Tina (June 19, 2012). "Subway Riders Want Rats Booted From Harlem Platform" (http://www.ny1.com/content/news_beats/transit/163355/subway-riders-want-rats-booted-from-harlem-platform). *NY1*. . Retrieved 2012-06-21.

[178] Namako, Tom (March 24, 2009). "You Dirty Rats! New Traps Target Subway Vermin" (http://www.nypost.com/seven/03242009/news/regionalnews/you_dirty_rats__new_traps_target_subway__161096.htm). *New York Post*. . Retrieved June 27, 2009.

[179] Redwine, Tina (October 5, 2011). "MTA To Launch New Cleaning Initiative To Combat Rat Infestations" (http://www.ny1.com/content/news_beats/transit/148410/mta-to-launch-new-cleaning-initiative-to-combat-rat-infestations). *NY1*. . Retrieved 2011-10-05.
[180] Epstein, Emily (October 24, 2011). "MTA removes bins hoping to reduce trash" (http://www.metro.us/newyork/local/article/1005728--mta-removes-bins-hoping-to-reduce-trash). *Metro*. . Retrieved 2011-12-13.
[181] "New York Subway Noise Levels Can Result In Hearing Loss For Daily Riders" *Science Daily* (http://www.sciencedaily.com/releases/2006/10/061012185519.htm) 2006-10-15
[182] "Got Ear Plugs? You May Want to Sport Them on the Subway and Other Mass Transit, Researchers Say" *Columbia Mailman School of Public Health News* (http://www.mailman.columbia.edu/news/article?article=761) 2009-06-18
[183] Bayen, Ann (March 29, 1976). "Token Women" (http://books.google.com/?id=Y-MCAAAAMBAJ&pg=PA46&lpg=PA46&dq=on+the+town+miss+subway&q=on the town miss subway). *New York Magazine* (New York Media, LLC): p. 46. .
[184] Klein, Alvin (June 6, 1993). "'On the Town' in Revival at [[Goodspeed Musicals|Goodspeed Opera (http://www.nytimes.com/1993/06/06/nyregion/theater-on-the-town-in-revival-at-goodspeed-opera.html?scp=6&sq=miss subway on the town&st=cse)]"]. *New York Times*.
[185] Ms. Subway 2004 Crowned (http://gothamist.com/2004/10/26/ms_subway_2004_crowned.php)
[186] MTA New York City Transit's Holiday Nostalgia Train (http://www.mta.info/mta/news/releases/?en=081125-NYCT188)
[187] The Vintage train runs every Saturday, November 26 to December 24 (http://www.mta.info/metrocard/promos/Vintagetrain/index.html)

External links

General information

- MTA NYC Transit Subways (http://mta.info/nyct/subway/index.html) (Official site with maps and schedules)
- Nycsubway.org (http://www.nycsubway.org/) (A thorough treatment of the current system and history, including historic and track maps)
- NYCT Subway Service@NYCTSubwayScoop (http://twitter.com/#!/NYCTSubwayScoop): The official New York City Subway Twitter web feed page
- NYC Subway on Facebook (http://www.facebook.com/pages/MTA-New-York-City-Transit/232635164606): The official New York City Subway Facebook page
- MTA Arts for Transit (http://www.mta.info/mta/aft/): The official MTA art guide
- MTAinfo (http://www.youtube.com/user/mtainfo): The official MTA YouTube web page
- MTAPhotos' Photostream (http://www.flickr.com/photos/mtaphotos): The official MTA Flickr web photo page
- New York Subway daily operations visualized (Java applet, unofficial; official subway map use licensed by MTA on 9/29/2011) (http://www.railbandit.com/sim-any.php?network=nycs)

Historical information

- MTA's own time line of events (http://www.mta.info/nyct/facts/ffhist.htm)
- Early Days of the NYC Subway (http://www.life.com/image/first/in-gallery/35272/early-days-of-the-nyc-subway) – slideshow by *Life magazine*
- The New York City Subway: A Century by John Stern (http://www.beautyofnyc.org/subwayhistory.htm)
- Abandoned Stations (http://www.columbia.edu/~brennan/abandoned/) (A thorough treatment of the history of the abandoned subway stations)
- The Subway Issue (http://www.nytimes.com/nyregion/nyregionspecial/) – special October 2010 regional news section by *The New York Times*
- "America's Newest Subway, First Built In Model", October 1931, Popular Science (http://books.google.com/books?id=CigDAAAAMBAJ&pg=PA61&dq=Popular+Mechanics+1931+curtiss&hl=en&ei=Kd0DTe6AD4KUnQeVpvDlDQ&sa=X&oi=book_result&ct=result&resnum=9&ved=0CEAQ6AEwCDgU#v=onepage&q&f=true)
- New York Subway Ride, 1905 (http://www.youtube.com/watch?v=xquZtKZDgcw)

BMT_West_End_Line

BMT West End Line	
The **D** train serves the entire BMT West End Line at all times	
Overview	
Type	Rapid transit
System	New York City Subway
Termini	Ninth Avenue Coney Island – Stillwell Avenue
Stations	13
Operation	
Opened	1916-1918
Owner	City of New York
Operator(s)	New York City Transit Authority
Character	Open Cut(9th Ave. Sta. only) Elevated
Technical	
No. of tracks	3-6
Track gauge	4 ft 8 $^1/_2$ in (1435 mm)
Electrification	600V DC third rail

The **BMT West End Line** is a line of the New York City Subway, serving the Brooklyn communities of Borough Park, New Utrecht, Bensonhurst, Bath Beach and Coney Island. The **D** train operates local on the entire line at all times and although there is a center express track and three express stations along the entire line, there is no regular express service at this time.

The elevated line, originally mapped as the **New Utrecht Avenue Line** (though the common name prevailed after construction), replaced the surface West End Line.

History

Under the Dual Contracts of 1913, an elevated line was built over New Utrecht Avenue, 86th Street and Stillwell Avenue, fully opening to Coney Island on July 21, 1917. The original surface right-of-way was retained for use by trolley cars to provide local service and protect the company's franchise. Over the years, the West End line has been featured in movies and television shows. The famous chase scene from "The French Connection" (1971) was filmed under the West End line in Brooklyn, and the opening scene of "Saturday Night Fever" (1977) features Tony Manero

(John Travolta) walking down 86th Street, with the West End elevated line above. Additionally, the opening credits of the television show "Welcome Back, Kotter" (1975) also featured the West End line.

Service patterns

Elevated line over 86th Street at Bay Parkway station

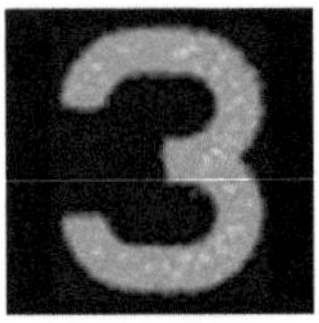

Route designation on BMT Triplex equipment

The West End Line has had an express (on the BMT Fourth Avenue Line) service - labeled 3 in 1924 - since it opened in 1916, passing over the Manhattan Bridge and onto the BMT Broadway Line express tracks. In the late 1950s, midday trains were switched to the local Fourth Avenue tracks and through the Montague Street Tunnel, and late night and Sunday service became a shuttle between Coney Island and 36th Street. The express and local services were assigned the designations **T** and **TT** in the early 1960s. With the opening of the Chrystie Street Connection in late 1967, the **B** train from Manhattan and the Bronx was extended to Coney Island, absorbing the T and TT (both ran express on Fourth Avenue). The **TT** late night and Sunday shuttle survived until 1968, when the **B** became full-time. It ran local on Fourth Avenue during late night hours, but express at all times. Late night operation was cut back to a shuttle to 36th Street in the mid-1980s. In 2001, when reconstruction of the Manhattan Bridge north tracks, the **B** service in Brooklyn was replaced by the new **W** train, which ran as a shuttle not only to 36th Street during nighttime hours, but also to Atlantic Avenue – Pacific Street on weekends. In 2002, reconstruction of Coney Island – Stillwell Avenue resulted in the West End Line being the only line to serve the terminal and the **W** was extended full-time into Manhattan, using the local Fourth Avenue tracks and Montague Street Tunnel on weekends and late nights hours. In 2004, the Manhattan Bridge reconstruction was completed, and the **W** was replaced with an extended **D** train, running over the bridge at all hours and express on Fourth Avenue except late nights.

The other service pattern was the "West End Short Line", a rush-hour local (on Fourth Avenue) service between the BMT Nassau Street Line in Lower Manhattan and 62nd Street or Bay Parkway. It became part of the TT in the early 1960s and was discontinued in 1967.

In 1987, the short line service was essentially recreated when the rush-hour **M** extension to Brooklyn was moved from the BMT Brighton Line to the West End Line terminating at Bay Parkway. It terminated at Ninth Avenue until 1995, when it was cut back to Chambers Street. It was extended again from 2001-04 while the Manhattan Bridge was closed for reconstruction. In 2010, as part of a series of MTA budget cuts, rush-hour **M** service was discontinued.

Chaining information

The West End Line is chained **BMT D**. This is coincidental to the fact that the current (2012) primary service designator is also **D**.

Station listing

Station service legend	
	Stops all times
	Stops rush hours only
Time period details	

Neighborhood	Station	Tracks	Services	Opened	Transfers and notes
splits from the BMT Fourth Avenue Line (D ●)					
Center Express track begins (No Regular Service)					
connecting tracks to former BMT Culver Line (No Regular Service)					
connecting tracks to 36th–38th Street Yard from local tracks					
Sunset Park	Ninth Avenue	all	D ●	June 24, 1916	former transfer to BMT Culver Line
Borough Park	Fort Hamilton Parkway	local	D ●	June 24, 1916	
	50th Street	local	D ●	June 24, 1916	
	55th Street	local	D ●	June 24, 1916	
Bensonhurst	62nd Street	all	D ●	September 15, 1916	BMT Sea Beach Line (N ●) at New Utrecht Avenue
	71st Street	local	D ●	September 15, 1916	
	79th Street	local	D ●	September 15, 1916	
	18th Avenue	local	D ●	September 15, 1916	
	20th Avenue	local	D ●	December 29, 1916	
	Bay Parkway	all	D ●	December 29, 1916	
	25th Avenue	local	D ●	December 29, 1916	
connecting track to Coney Island Yard					
Gravesend	Bay 50th Street	local	D ●	December 21, 1917	
connecting track to Coney Island Yard					
Center Express track ends					
Coney Island	Coney Island – Stillwell Avenue	all	D ●	December 23, 1918	BMT Brighton Line (Q ●) IND Culver Line (F ●) BMT Sea Beach Line (N ●)

See also

- Transportation to Coney Island

References

- Opening of the Brooklyn, Bath and Coney Island Railroad, New York Times June 9, 1864 page 2
- Railroad History Database [1]

External links

- Gunther and His Railroad by Morton Morris [2]
- BMT West End Line on NYCSubway.org [3]

References

[1] http://www.earlpleasants.com/search_1.asp
[2] http://www.3drail.com/9909/westend1.htm
[3] http://nycsubway.org/lines/westend.html

Second_Avenue_Subway

Second Avenue Subway

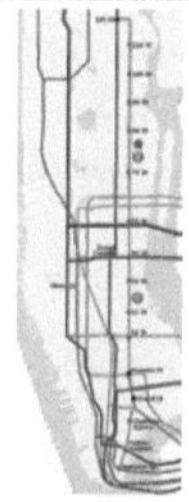

Planned route of the NYC Subway Second Avenue Line ("T", in teal).

Overview	
Type	Rapid transit
System	New York City Subway
Status	Under construction
Locale	Manhattan, New York City, USA
Termini	125th Street Hanover Square
Stations	16
Operation	
Opened	First Phase: Late 2016 (expected)
Owner	City of New York
Operator(s)	New York City Transit Authority / MTA Capital Construction
Technical	
Line length	8.5 mi (km)
No. of tracks	2
Track gauge	4 ft 8 $\frac{1}{2}$ in (1435 mm)
Electrification	600 V DC third rail

The **Second Avenue Subway (SAS)** is a planned rapid transit subway line, part of the New York City Subway system. As of 2012, Phase I, consisting of two miles (km) of tunnel and three stations, is under construction underneath Second Avenue in the borough of Manhattan.

The Second Avenue Subway has been a plan and occasional construction project since 1929. After several starts and interruptions, mostly because of lack of funds, the most recent and financially secure construction plan was launched when a tunnelling contract was awarded to the consortium of Schiavone/Shea/Skanska by the Metropolitan Transportation Authority (MTA) on March 20, 2007.[1] This followed preliminary engineering and a final tunnel design completed by a joint venture between AECOM and ARUP.[2] [3] This contract, and the full funding grant agreement with the Federal Transit Administration which was received in November 2007, is for Phase I of the project, a newly built line between the existing BMT 63rd Street Line and 96th Street and Second Avenue.[4] The total cost of the 8.5-mile (km) line is expected to be over $17 billion.[5] Parsons Brinckerhoff is currently serving as the Construction Manager of the project.

A ceremonial ground-breaking for the Second Avenue Subway was held on April 12, 2007 and the contractor prepared the initial construction site at 96th Street on April 23, 2007. A tunnel boring machine (TBM) was originally expected to arrive six to eight months after construction began, but the utility relocation and excavation required to create its "launch box" delayed its deployment until May 2010.[4] By May 2010 the TBM launch box was complete, and on May 14, 2010, MTA's contractors completed the TBM installation and turned it on.[6] [7] [8] On March 28, 2011, S3, having completed the west tunnel to 65th Street, drilling began for the east tunnel. The TBM completed its run to the 63rd Street Station bellmouth on September 22, 2011.

The reasons for the line's many "false starts" and delays are numerous and complex. The line is sometimes referred to as "The Line That Time Forgot".[5] [9]

Background

Originally proposed in 1929 as part of a massive expansion of the Independent Subway System (IND), work on the line never commenced, as the Great Depression crushed the economy of the state and country. Need for the Second Avenue Subway line grew, especially in recent years, as the East Side of Manhattan experienced significant residential development. Currently, the lone rapid transit option on the Upper East Side is the four-track IRT Lexington Avenue Line, the most crowded in the country. Its average of 1.3 million daily riders on this single line exceeds the daily traffic load of the entire Washington Metro system (which has the second-highest ridership in the U.S.), and exceeds the *combined* daily ridership of the rail transit systems of San Francisco and Boston.[10] Local bus routes are just as crowded during various times of the day. The construction of the Second Avenue line would add another two tracks to fill the gap that has existed since the elevated IRT Second Avenue Line was demolished in 1940–42 and the IRT Third Avenue Line was removed in 1955–56.

The city started planning, again, in 1945, to build the new subway and bought a prototype train (the R11) in 1949 for use on the new line.[11] New York voters approved bond acts for its construction in 1951 and in 1967. Money from the 1951 bond measure was diverted to buy new cars, lengthen platforms, and maintain other parts of the aging New York City subway system. The proceeds of the 1967 bond act were partly used to begin tunneling under Second Avenue. Digging began in 1972; however, a few years later, the city became insolvent. "It's the most famous thing that's never been built in New York City, so everyone is skeptical and rightly so," said Gene Russianoff, an advocate for subway riders since 1981. "It's much-promised and never delivered."

On November 8, 2005, voters in New York State passed the Transportation Bond Act, which will, among other projects, partially fund construction of the line. Its passage had been seen as critical to its construction. After warning that failure to pass the act would doom the project, MTA chairman Peter S. Kalikow stated that "Now it's up to us to complete the job" given its approval by a 55–45 percent margin.[12]

In August 2006, the MTA revealed that all future subway stations, including ones built for the Second Avenue subway, the 7 Subway Extension, and the new South Ferry station will be outfitted with air-cooling systems to reduce the temperature along platforms.[13]

In November 2007, Mary Peters, the United States Secretary of Transportation announced that the Second Avenue Subway would receive $1.3 billion in federal funding for the project's first phase, to be funded over a seven-year period.[14]

History

The Second Avenue El, looking south on First Avenue from 13th Street during its demolition in September 1942

The need for a subway line under Manhattan's Second Avenue was realized shortly after the First World War. In 1919, the New York Public Service Commission launched a study at the behest of engineer Daniel L. Turner to determine what improvements were needed in the city's public transport system. The Second Avenue Elevated operated above Second Avenue north of the Queensboro Bridge until 1940, and south to downtown, part of the way on First Avenue, until June 13, 1942.[15] The Third Avenue Elevated operated a block to the west until 1955.

Turner's final paper, titled *Proposed Comprehensive Rapid Transit System*, was a massive plan calling for new routes under almost every north-south Manhattan avenue, extensions to lines in Brooklyn and Queens, and several crossings of The Narrows to Staten Island. Massively scaled-down versions of some of Turner's plans were found in proposals for the new city-owned Independent Subway System (IND). Among the plans was a massive trunk line under Second Avenue consisting of at least six tracks and numerous branches throughout Brooklyn, Queens, and the Bronx.

In 1929, the Board of Transportation of the City of New York tentatively approved the expansion, which included a Second Avenue Line with a projected construction cost of $98,900,000, not counting land acquisition. From north to south, the 1929 plan included four tracks from the Harlem River (where it would continue north as a Bronx trunk line with several branches) to 125th Street, six tracks from 125th Street to a link with the IND Sixth Avenue Line at 61st Street, four tracks from 61st Street to Chambers Street, and two tracks from Chambers Street to Pine Street.

Due to the Great Depression, the soaring costs of the expansion became unmanageable. Construction on the first phase of the IND was already behind schedule, and the city and state were no longer able to provide funding. A scaled-down proposal including a turnoff at 34th Street and a connection crosstown was postponed in 1931.

Further revision of the plan and more studies followed. By 1939, construction had been postponed indefinitely, and Second Avenue was relegated to "proposed" status. The 1939 plan for subway expansion took the line not only into the Bronx (by now as a single line to Throggs Neck) but also south into Brooklyn, connecting to the stub of the IND Fulton Street Line at Court Street.

The United States' entry into World War II in 1941 halted all but the most urgent public works projects, delaying the Second Avenue Line once again.

Post-war

Finally, in 1945, plans for the Second Avenue Subway were again revised. The southern two-track portion was abandoned as a possible future plan for connecting the line to Brooklyn.

A 1947 plan once again connected the Second Avenue Line to Brooklyn, but via the BMT trackage over the Manhattan and Williamsburg Bridges. A connection would allow trains from these bridges to go onto the IND Sixth Avenue Line rather than the Second Avenue Line. Other connections to the Second Avenue Line were to be provided at 57th Street, via a line connecting to the Sixth Avenue Line; two express tracks would be built along that line north of West Fourth Street. The IRT Pelham Line would be switched to the combined IND/BMT division (this plan also includes other connections, which have been built), and connected to the Second Avenue Line. The Second Avenue Line would end just north of that connection, at 149th Street, with transfers to the IRT White Plains Road Line and the elevated IRT Third Avenue Line, the latter of which would be demolished south of 149th Street.

In 1949, the New York Board of Transportation accepted delivery of ten new prototype subway cars made of stainless steel from the Budd Company, named by their contract, R11, specifically intended for the Second Avenue Subway. They cost $100,000 each; the train became known as the "million dollar train". The cars featured porthole style round windows and a new public address system. Reflecting public health concerns of the day, especially regarding polio, the R11 cars were equipped with electrostatic air filters and ultraviolet lamps in their ventilation systems to kill germs.[11] [16]

By 1950, the plans called for a connection from Second Avenue at 76th Street to 34th Avenue in Queens, via a new tunnel under the East River. The city was barely able to raise money for the construction effort, but the onset of the Korean War caused soaring prices for construction materials and saw the beginning of massive inflation.

A 1954 plan added another feeder, an East River tunnel at 76th Street, connecting existing Long Island Rail Road (LIRR) trackage (which would be converted for subway use) to the Second Avenue Line towards downtown. This plan has been revitalized as part of the 2005 Transportation Bond Act, which would connect the LIRR trackage to Grand Central Terminal via the 63rd Street Tunnel as part of the East Side Access project.

The southernmost part of the 1947 plan, connecting the two BMT bridges to the IND Sixth Avenue Line, was built in the 1960s and opened in 1967 as the Chrystie Street Connection. Other parts of that plan were carried out, including the connection at 57th Street (moved to 63rd Street) and the abandonment of the IRT Third Avenue Line south of 149th Street, but the rest of the Second Avenue Line was not built. Plans now call for an additional two tracks in the Chrystie Street area for the Second Avenue mainline; current plans have the new tracks under the old ones, while older plans had one track on each side of the Chrystie Street Connection.

1970s: Completed segments

In 1964, Congress passed the Urban Mass Transportation Act, promising federal money to fund mass transit projects in America's cities via the Urban Mass Transportation Administration. In 1967, voters approved a $2.5 billion Transportation Bond Issue, which provided over $600 million for New York City projects. The Second Avenue project was given top priority, and would stretch from 34th Street to The Bronx. The City secured a UMTA grant for initial construction, and a groundbreaking ceremony was held on October 27, 1972. Construction began shortly thereafter at Second Avenue and 103rd Street.

However, the city soon experienced its most dire fiscal crisis yet. The stagnant economy of 1975, combined with the massive outflow of city residents to the suburbs, led to a fiscal disaster for the city. Construction of the subway was halted, with only three sections of tunnel having been completed, in addition to the Chrystie Street Connection. These sections are between Pell and Canal Street, 99th and 105th, and 110th and 120th Streets. The two northern sections between 99th and 105th, and 110th and 120th Streets, will be used in Phase 2 of the current SAS plan (96th to 125th). The section from Pell to Canal will not be used under the current preferred alternative, which will bring the line a few blocks away from this section. Construction was also begun between Second and Ninth Streets, though the extent is unknown; some rumors say that only utilities were relocated, while others say that it was excavated but filled back in.[17]

Current development

With the city's economic and budgetary recovery in the 1990s, there was a revival of efforts to complete construction of the SAS. Rising ridership on the IRT Lexington Avenue Line, the only subway trunk line east of Central Park in Manhattan, demonstrated the need for the Second Avenue Line, as capacity and safety concerns rose.

The MTA's final environmental impact statement was approved in April 2004; the latest proposal is for a two-track line from 125th Street and Lexington Avenue in Harlem, down Second Avenue to Hanover Square in the Financial District. The new subway line will actually carry two services. The full-length Second Avenue line, extending from Harlem to the financial district, probably will be given the teal **T** as its letter designation. However, it is the other service, a proposed reroute of the **Q**, which will begin carrying passengers first.

The MTA proposes to build the Second Avenue Subway in four segments with connections to other subway lines. The first segment (Phase I) is a proposed reroute of the **Q**, BMT Broadway Line across 63rd Street and north along Second Avenue to the Upper East Side at 96th Street. The other three segments, in the order that they are proposed to be built, are a reroute of the **Q** train to 125th Street from 96th Street (Phase II), 63rd Street to Houston Street (Phase III, introduction of the **T** train), and Houston Street to Hanover Square, Manhattan (Phase IV, full length **T** train service).[18]

Computer-generated image of a future Second Avenue Subway station

New York voters passed a transportation bond issue in November 2005, allowing for dedicated funding allocated for Phase I; the first construction contract of the current plan finally signed, the MTA is expected to receive a funding agreement from the federal government to complete Phase I.

The subway will be built with deep bore tunneling methods, avoiding the cumbersome utility relocation and cut-and-cover methods of past generations that made subway building disruptive for traffic, pedestrians, and store owners. Stations will retain cut-and-cover construction.[19]

Construction began with utility relocation; the MTA anticipated completing this step in eight months, but was not finished 14 months after commencement. For boring, a trench will be dug from 96th to 93rd Streets; the tunnel boring machine will be placed in the ground at 92nd Street and will bore southbound, connecting shafts at 86th and 72nd Streets, which to be sunk as starting points for subway stations. Tunneling is expected to take approximately one year.[20] As of June 2008, substantial portions of the utility relocation work between 91 St. and 96 St. were completed.

The U.S. Department of Transportation announced on December 18, 2006, they would allow the MTA to commit up to $693 million in funds to begin construction of the Second Avenue Subway Line and that the federal share of such costs would be reimbursed with FTA transit funds, subject to appropriations and final labor certification.[21]

In late January 2007, the media reported that a $333 million contract would be awarded within weeks to three American firms to build Phase One. The actual price was $337 million; the TBM will begin at 92nd Street, not 96th Street as reported. The station site at 96th Street will see cut and cover construction.[22] [23] [24] [25]

Groundbreaking for the Second Avenue Subway construction project was on April 12, 2007, in a tunnel segment built in the 1970s at 99th Street. The MTA reported that the 1970s Second Avenue subway tunnel (which will be part of Phase I and Phase II) is in pristine condition.[26]

In June 2008, the MTA, facing cost increases for construction materials and diesel fuel affecting the prices of contracts not yet signed, announced that certain features of the Second Avenue Subway would be simplified to save money. One set of changes, which significantly reduces the footprint of the subway in the vicinity of 72nd Street is the alteration of the 72nd Street Station from a three-track, two-platform design to a two-track, single island platform design, paired with the elimination of the third track and simplification of the planned interlocking merging the BMT Broadway Line's Second Ave extension and the planned **T** subway, extending south of 63rd Street along Second Avenue. Supplemental environmental impact studies covering station configuration options for the proposed 72nd Street and 86th Street stations are underway.[27] [28]

Construction methods

Planned construction methods vary depending on the section of the line, due to varying underground conditions. The methods planned for each section are as follows:[29] [30]

Streets	Construction method	Streets	Construction method	Streets	Construction method	Streets	Construction method	Streets	Construction method
125-120	Tunnel Boring Machine	106-101	Existing	73-71	Mined with Cut and Cover	41-34	Tunnel Boring Machine	15-13	Cut and Cover
120-117	Existing	101-95	Cut and Cover	71-58	Tunnel Boring Machine	34-32	Cut and Cover		
117-114	Cut and Cover	95-87	Tunnel Boring Machine	58-56	Cut and Cover	34-24	Tunnel Boring Machine		
114–109	Existing	87-84	Mined with Cut and Cover	56-43	Tunnel Boring Machine	24-22	Mined with Cut and Cover		
109-106	Cut and Cover	84-73	Tunnel Boring Machine	43-41	Mined with Cut and Cover	22-15	Tunnel Boring Machine		

A number of different methods will be used to tunnel for 13.7 kilometers (mi) underneath Manhattan, which is densely populated. 90% of the tunneling will be performed by a tunnel boring machine. The rest of it will be done using the cut and cover method and mined drill and blast, for sections, generally the 16 stations, that average 275 meters (ft) in length. The stations at 86th and 72nd Streets will be mined. This will be challenging, given the number of high value, high rise properties in their vicinities. The 96th Street cut and cover station will be at about 15 meters (ft) deep, making it one of the shallowest stations. Stations at the two mined stations will be between 25.9 meters (ft) and 27.4 meters (ft) deep in rock. The construction method that will be used should ease concerns for the above buildings, because only two shafts will be required for excavation.[31]

In Phase 1 there will be tunneling between East 63rd and 92nd Streets and a 248-meter-long (ft) by 23-meter-wide (ft) TBM launch box will be built. That will ultimately become part of the 15-meter-deep (ft) 96th Street Station. Two access shafts will be constructed for the East 72nd Street Station. Slurry or diaphragm walls, 1.1 meters (ft) wide and 6.1 meters (ft) long and approximately 35 meters (ft) deep, will be built alongside the sections between

East 93rd and 95th Streets. Inasmuch as the rock is shallower between East 91st and 93rd Streets, 1.1-meter-diameter (ft) secant piles will do the same work at shallower depths.[31]

Earth excavation will be conducted between walls, once they are installed, and a box structure will be built using a bottom-up construction method. Temporary decking will constitute the top of the box, and the decking will both brace the excavation and support the walls and Second Avenue traffic.[31]

72nd Street station cavern as of January 2012

The tunnels and stations will be up to approximately 30 meters (ft) below street level. Of the below-ground obstacles, Arup director of construction David Caiden says: "It's a spaghetti of tunnels, utilities, pipes and cables – I've never seen anything like it."[31] Additionally, the project must go over, or under, subway lines, Amtrak railway lines, and the Queens-Midtown Tunnel linking Manhattan and Queens.[31]

Tunnel at 64th Street

There are geological anomalies along the way. Manhattan's geology changes along the subway's length, passing through rock and soft ground, consisting of sands, silts, and clays over Manhattan schist, and there are faults and shear zones as well as fractured rock. Hard-rock Tunnel Boring Machines 6.7 meters (ft) in diameter will tunnel during the first phase, progressing at anticipated rate of about 20 meters (ft) per day.[31]

Planned SAS route/stations and designation

83rd Street

The plans for the Second Avenue Subway involve digging 8.5 miles (km) of new tunnel from 125th Street in Harlem south to Hanover Square, which is located in Manhattan's Financial District. Initially, during Phase I, the line will begin at the intersection of Second Avenue and 96th Street, running south to join the BMT Broadway Line via the existing BMT 63rd Street Line. Phase I stations will be located at 96th Street, 86th Street and 72nd Street. Plans call for the **Q** service to be re-routed to 96th Street (though the effect of the 2010 subway service changes on these plans are unknown), and then in Phase II to 125th Street and Lexington Avenue. After Phase III, the new[32] **T** service will operate from 125th Street to Houston Street. After Phase IV opens, **T** service will run the full length of the line, from 125th Street to Hanover Square.

The MTA decided to designate the main, full-length Second Avenue service the **T** in part for the following reasons:

- The letters **O** and **I** are too easily confused with the digits **0** and **1**, respectively.[33]
- The letters **H** and **K** were used in the relatively recent past to denote services on the IND Eighth Avenue Line, which includes the **A**, **C** and **E** trains, and thus are not preferred.
- The letters *P*, *U* and *Y* are too similar to words.[33]

This left the letter **T** as the final choice.[34]

The new stations of the completed Second Avenue Line are proposed as follows, with Phase 1 under construction:

Station	Phase	Transfers & Notes
125th Street	2	Northern terminal station for **Q** train (Phase 2) and **T** train (Phase 3) 4 5 6 <6> trains (IRT Lexington Avenue Line) M60 bus to LaGuardia Airport connection to Harlem – 125th Street (Metro-North Railroad) at Lexington Avenue and 125th Street
Provision for expansion to The Bronx		
116th Street	2	
106th Street	2	
96th Street	1	Northern terminal station for **Q** train in Phase 1
86th Street	1	
72nd Street	1	
Q train splits to BMT Broadway Line via BMT 63rd Street Line (Phase 1)		
T train continues down Second Avenue (Phase 3)		
55th Street	3	E M trains (IND Queens Boulevard Line) at Lexington Avenue – 53rd Street 4 6 <6> trains (IRT Lexington Avenue Line) at 51st Street
42nd Street	3	7 <7> trains (IRT Flushing Line) S train (IRT 42nd Street Shuttle) 4 5 6 <6> trains (IRT Lexington Avenue Line) at Grand Central – 42nd Street connection to Grand Central Terminal (Metro-North Railroad & Long Island Rail Road once East Side Access Project is completed.)
34th Street	3	
23rd Street	3	
14th Street	3	L train (BMT Canarsie Line) at Third Avenue
Houston Street	3	F train (IND Sixth Avenue Line) at Second Avenue Southern terminal station for **T** train (Phase 3)
Grand Street	4	B D trains (IND Sixth Avenue Line) at Grand Street
Chatham Square	4	at Worth Street
Seaport	4	at Fulton Street
Hanover Square	4	at Old Slip Southern terminal station for **T** train (Phase 4)
Provision for expansion to Brooklyn		

The above stations will serve the Second Avenue main service, terminating at 125th Street and at Hanover Square. In addition to the main service, tentatively dubbed the **T**, and colored turquoise, a connection is planned to the BMT Broadway Line, utilizing an existing connection via the BMT 63rd Street Line, as part of phase 1. It is likely that the **Q** service will be extended northward from 57th Street – Seventh Avenue, curving east under Central Park on the unused portion of the BMT 63rd Street Line. The **Q** train would stop at Lexington Avenue – 63rd Street with a cross-platform interchange to the IND 63rd Street Line (**F** train) before merging with the Second Avenue Line at 64th Street. Thus, the residents of Spanish Harlem and the Upper East Side will have direct mass transit service down both Second Avenue and Broadway to the Financial District, and across the Manhattan Bridge to Brooklyn via the **Q** train.

An additional two-track connection is planned between the line towards Lower Manhattan (around 62nd Street) and the IND 63rd Street Line towards Queens; current plans don't call for it to be used by regular service. Provisions are also being made for an extension north under Second Avenue past 125th Street to the Bronx, and an extension south to Brooklyn. No track connection will be provided to the Chrystie Street Connection.

Just north of Broome Street, the subway will pass under a short unused highway tunnel, the only part of the Lower Manhattan Expressway to be built.

Construction status

In March 2007, the MTA awarded a contract for constructing the tunnels between 92nd and 63rd Streets, a launch box for the tunnel boring machine (TBM) at 92nd to 95th Streets, and access shafts at 69th and 72nd Streets. This contract, valued at $337 million, was awarded to a joint venture of Schiavone Construction, Skanska USA Civil and San Francisco-based J.F. Shea Construction.

The ceremonial groundbreaking for the first phase of the Second Avenue subway was held on April 12, 2007.[35] Actual construction work began, on the surface of Second Avenue between 91st and 95th Streets, on April 23, 2007.

On May 28, 2009, the MTA awarded a $303.8 million contract to E.E. Cruz and Tully Construction Co., a Joint Venture and LLC, to construct the 96th Street station box.[36] Work began in July on site clearing and utility relocation necessary to prepare for the installation of slurry walls between 95th and 99th Streets where the station connects to the existing tunnel section built in the 1970s.

In June 2009, the first of three contracts for the 86th Street Station was awarded for the advance utility relocation work and construction of cut and cover shaft areas at 83rd and 86th Streets. This contract provided two vertical starter shafts that will be used by a subsequent contractor to excavate the station cavern in the rock between 83rd and 86th Streets.

On June 5, 2009 an apartment building at 1772 Second Avenue was evacuated by the NYC Department of Buildings (DOB) after it was determined that the building was in danger of collapse.[37] Then on June 29, 2009 the DOB evacuated a mixed use building at 1768 Second Avenue/301 East 92nd Street because it too was in danger of collapse.[38] The evacuation of these two buildings has delayed the contractor's plan to use controlled blasting to remove bedrock in the southern section of the launch box.[39] Until the blasting permits could be issued, MTA required contractors to use mechanical equipment to remove the bedrock, which is slower than blasting out the rock.[40] As of late October, 2009, one building has been shored up, and work is in progress on the second; MTA has rescheduled blasting to begin during the week of November 2.[41]

On October 1, 2010, MTA awarded a $431 million contract to SSK Constructors (a joint venture) for the mining of the tunnels connecting the 72nd St station to the existing 63rd Street station, and for the excavation and heavy civil structures of the 72nd St Station. Subsequent contracts will be awarded for the following: excavation of the cavern at 86th Street Station; architectural and mechanical and electrical work at 72nd, 86th and 96th Street Stations; rehabilitation of the 63rd Street Station; and the Systems Contract (track, signals and communications) for the entire Phase 1 area. On January, MTA awarded Judlau Contracting a 40-month, $176,400,000 contract to rebuild and enlarge the Lexington Ave/63rd Street Station.[42]

The MTA and its contractors on the project meet on a regular basis with the Manhattan Community Board 8 Second Avenue Subway Task Force and Manhattan Community Board 11 to report on construction progress and to seek input from the community.[43] The MTA maintains a Construction Look Ahead [44] web page that summarizes the planned construction activity for the next three weeks.

On May 14, 2010, the Tunnel Boring Machine was started at the Second Avenue Subway launch box at 96th Street. The TBM dug at a rate of approximately 50 feet per day; the TBM finished its run at the planned endpoint under 65th Street on February 5, 2011.[45] S3 partially disassembled the TBM and backed it out of the tunnel. It was repositioned in the east starter tunnel to begin boring again.[46] Because the east side of Second Avenue has some soft ground not compatible with the Robbins TBM, ground-freezing was undertaken to prepare the soil for the TBM.[7] [47] [48] On March 28, 2011, S3, having completed its task of completing the 7200-foot (m) west tunnel to 65th Street, began drilling the east tunnel, with the first 200 feet (m) being through soil frozen by S3 using calcium chloride fed through a network of pipes.[49] The TBM drilling the east tunnel will negotiate the curve onto 63rd Street and break through the bellmouth at the existing 63rd St subway station. The portion of the west tunnel remaining to be created will be mined using conventional drill-and-blast methods, because the curve S3 construction teams would have to negotiate is too tight for the TBM.

On September 22, 2011, the TBM completed its run to the 63rd Street/Lexington Av Station bellmouth.

Workers celebrate after the TBM reaches the BMT 63rd Street Line.

Estimated completion schedules have provided much fodder for critics. The most recent proposed construction schedule has the Second Avenue Subway open for passenger service in 2016.[50] Other recent publications have listed expected construction dates as follows:

- 2007–16:[51] Phase 1 (96th St. to 63rd St.) State Funding In-Place, Federal Funding Approved.[52] In its 2008 capital improvement budget proposal, the MTA pushed back completion of Phase 1 from 2014 to 2015. In 2009 the MTA pushed it back again to 2016.[53]
- Phase 2 (125th St. to 96th St.) No funding commitments.
- Phase 3 (63rd St. to Houston St.) No funding commitments.
- Phase 4 (Houston St. to Hanover Square) No funding commitments.

See also

- BMT 63rd Street Line
- Chrystie Street Connection
- East Side Access
- MTA Capital Construction
- Transportation in New York City

References

[1] MTA Press Release March 20, 2007

[2] Arup to design new Second Avenue Subway, New York | Arup (http://www.arup.com/americas/newsitem.cfm?pageid=873)

[3] Second Avenue Subway | Transport Consulting | Arup (http://www.arup.com/transportconsulting/project.cfm?pageid=7267)

[4] "Second Avenue Subway – A Status Report" (http://www.arema.org/eseries/scriptcontent/custom/e_arema/library/2008_Conference_Proceedings/Second_Avenue_Subway-A_Status_Report_2008.pdf) (PDF). . Retrieved August 2, 2009.

[5] Sargent, Greg (March 29, 2004). *The Line That Time Forgot – Second Avenue Subway* (http://nymag.com/nymetro/news/features/n_10109/). Nymag.com. . Retrieved August 2, 2009.

[6] " "Blasting on Second Avenue" (http://thelaunchbox.blogspot.com/2010/03/march-19-20-2010-blasting-on-second.html). thelaunchbox.blogspot.com. March 22, 2010. ".

[7] Siff, Andrew (May 14, 2010). "2nd Ave. Subway Tunnel Dig Begins" (http://www.nbcnewyork.com/news/local-beat/2nd-Ave-Subway-Tunnel-Dig-Begins-93787364.html). *WNBC*. . Retrieved May 14, 2010.

[8] "Tunneling Begins Under Second Avenue" (http://www.mta.info/news/stories/?story=63). *MTA*. May 14, 2010. . Retrieved May 17, 2010.

[9] "The line that time forgot" (http://www.plannyc.org/node/31960). PlanNYC. April 5, 2004. . Retrieved August 2, 2009.

[10] "Second Avenue Subway in the Borough of Manhattan, New York County, New York Final Environmental Impact Statement And Final Section 4(f) and Section 6(f) Evaluation" (http://www.mta.info/capconstr/sas/feis.htm) (PDF). 2004-04. pp. 1-5, 1-6. . Retrieved 2011-09-22.

[11] Neuman, William (March 24, 2007). "A Museum-Quality Car for a Subway Yet Unbuilt" (http://www.nytimes.com/2007/03/24/nyregion/24subway.html). *New York Times*. . Retrieved February 20, 2010.

[12] Chan, Sewell (November 9, 2005). "Voters Approve Transit Bonds for $2.9 Billion" (http://www.nytimes.com/2005/11/09/nyregion/metrocampaigns/09transport.html). *The New York Times*. . Retrieved September 20, 2007.

[13] Donohue, Pete (August 4, 2006). "Cooler Subways Coming Eventually" (http://web.archive.org/web/20071010000252/http://www.nydailynews.com/news/2006/08/04/2006-08-04_cooler_subways_coming__eventually.html). *Daily News* (New York). Archived from the original (http://www.nydailynews.com/news/2006/08/04/2006-08-04_cooler_subways_coming__eventually.html) on October 10, 2007. . Retrieved December 12, 2008.

[14] Neuman, William (November 19, 2007). "U.S. Approves $1.3 Billion for 2nd Avenue Subway" (http://www.nytimes.com/2007/11/19/nyregion/19subway.html). *The New York Times*. . Retrieved November 19, 2007.

[15] Staff. "Second Avenue 'El' Coming to a Stop" (http://pqasb.pqarchiver.com/csmonitor_historic/access/278313072.html?dids=278313072:278313072&FMT=ABS&FMTS=ABS:AI&date=Jun+13,+1942&author=By+a+Staff+Correspondent+of+The+Christian+Science+Monitor&pub=Christian+Science+Monitor&desc=Second+Avenue+'El'+Coming+to+a+Stop&pqatl=google), *The Christian Science Monitor*, June 13, 1942. Retrieved October 12, 2008.

[16] R-11 Datasheet

[17] FAQ: Completed Portions of the 2nd Avenue Subway (http://www.nycsubway.org/ind/2ndave/builtfaq.html). Retrieved August 4, 2006.

[18] Construction phasing

[19] Stabile, Tom (May 2006). "New York's Subway System Finally Starting Major Expansion (http://web.archive.org/web/20080509014513/http://construction.com/NewsCenter/Headlines/RP/20060523ny.asp)"]. newyork.construction.com. Archived from the original (http://construction.com/NewsCenter/Headlines/RP/20060523ny.asp) on May 9, 2008. . Retrieved December 12, 2008.

[20] 2nd Ave subway could get early start (http://www.amny.com/news/local/transportation/subway/am-mta1024,0,4756878.story), *AM New York*, October 24, 2006.

[21] "U.S. Transportation Secretary Signs Record $2.6 Billion Agreement to Fund New Tunnel Network To Give Long Island Commuters Direct Access to Grand Central Station" (http://www.dot.gov/affairs/dot11706.htm), United States Department of Transportation press release dated December 18, 2006. Retrieved September 20, 2007.

[22] Exclusive: Ground Breaking For 2nd Avenue Subway Line Weeks Away (http://www.ny1.com:80/ny1/content/index.jsp?stid=8&aid=66187) – NY1, January 24, 2007

[23] 2nd Avenue Subway Contract Signed (http://www.wnyc.org/news/articles/75760) – WNYC Newsroom, March 21, 2007

[24] Olshan, Jeremy (March 21, 2007). "Second Ave. Tunnel Vision" (http://www.nypost.com/seven/03212007/news/regionalnews/second_ave_tunnel_vision_regionalnews_jeremy_olshan.htm). *New York Post*. . Retrieved January 9, 2009.

[25] "MTA Signs Second Ave. Subway Contract" (http://www.nysun.com/new-york/mta-signs-second-ave-subway-contract/50892/). *New York Sun*. March 21, 2007. . Retrieved February 20, 2010.

[26] Date set for 2nd Ave line groundbreaking (http://www.amny.com/news/local/am-subway0329,0,3016869.story?coll=am-local-headlines) – AmNY, March 28, 2007

[27] "Microsoft PowerPoint – 0080617_CB8_Final [Compatibility Mode]" (http://www.mta.info/capconstr/sas/documents/CB8 presentation 06-17-08.pdf). . Retrieved August 2, 2009.

[28] "Microsoft PowerPoint – 080729_CB8_Final_distribution version" (http://www.mta.info/capconstr/sas/documents/cb8_presentation.pdf). . Retrieved August 2, 2009.

[29] Second Avenue Subway (http://www.mta.info/capconstr/sas/documents/2ndave.pdf)

[30] Photograph of 95th St – 101 Street Tunnel Contract Notice posted by EE Cruz-Tully Joint Venture (http://2.bp.blogspot.com/_tjgTAe0iBhA/Sp3GSm2oX8I/AAAAAAAACcQ/NCfRSeoDXYw/s1600-h/IMG_9776c.JPG)

[31] Wynne, Alexandra (January 20, 2009). "Fairytale of New York – Second Avenue Subway takes shape | Features | New Civil Engineer" (http://www.nce.co.uk/fairytale-of-new-york-second-avenue-subway-takes-shape/1970428.article). Nce.co.uk. . Retrieved August 2, 2009.

[32] Hirschman, David (August 2008). "The T Train: NYC Will Get Its First New Subway Line in 70 Years" (http://www.wired.com/cars/futuretransport/magazine/16-08/st_atlas). *Wired.com* (Wired Magazine) (Aug '08): 36. . Retrieved 2012-08-04. "The old (1960s) T service was also called the West End train. The reference was to Brooklyn. By contrast, the new T service will serve the East Side of Manhattan, and "will unite the Upper and Lower East Sides.""

[33] V and W Trains join a long list of routes that have bowed out of the Subways (http://amny.com/urbanite-1.812039/v-and-w-trains-join-a-long-list-of-routes-that-have-bowed-out-of-the-subways-1.2028944), AM New York. Accessed 2012-09-06

[34] Reeves, Hope (October 26, 2006). "The Second Avenue Subway Is Brought to You by the Letter T" (http://nymag.com/daily/intel/2006/10/take_my_ttrain_please.html). *New York Magazine*. . Retrieved August 2, 2009.

[35] "Second Avenue subway groundbreaking: Is 4th time the charm?" (http://www.thejournalnews.com/apps/pbcs.dll/article?AID=/20070412/NEWS05/704120441). *The Journal News*. April 12, 2007. . Retrieved April 18, 2007.

[36] "MTA Capital Construction Recent Contract Awards, retrieved May 30, 2009" (http://www.mta.info/capconstr/procurement/cc_recentawards.htm). Mta.info. . Retrieved August 2, 2009.

[37] Namako, Tom (June 6, 2009). "2nd Ave. Subway Caused Building Evac: Officials" (http://www.nypost.com/p/news/regional/manhattan/nd_ave_subway_caused_building_evac_cJn5kslNXhmU3gF509BQaK). *New York Post*. .

[38] Sutherland, Amber; Namako, Tom (July 1, 2009). "Second Ave. Tenants RIP 'Train Wreck'" (http://www.nypost.com/p/news/regional/nd_ave_tenants_rip_train_wreck_1bOFrUdCgHGb1uVcxoTL6L). *New York Post*. .

[39] Rivoli, Dan (September 2, 2009). "2nd Ave. Subway Delays" (http://ourtownny.com/?p=4058). Our Town. . Retrieved February 20, 2010.

[40] Simeone, Jessica; Namako, Tom (September 26, 2009). "Second Ave. on Snail Rail" (http://www.nypost.com/p/news/local/manhattan/second_ave_on_snail_rail_pPG3DCDAARkGv19J9OihzH). *New York Post*. .
[41] Second Avenue Subway (http://www.mta.info/capconstr/sas/construction.html#blast)
[42] MTACC Procurement (http://www.mta.info/capconstr/procurement/cc_recentawards.htm)
[43] "Microsoft PowerPoint – 090323 S3 Construction Update _final [Compatibility Mode]" (http://www.mta.info/capconstr/sas/documents/090323_CB8 Combined presentation.pdf) (PDF). . Retrieved August 2, 2009.
[44] http://www.mta.info/capconstr/sas/construction.html
[45] COMMUNITY BOARD EIGHT PROJECT UPDATE, MTACC, February 15, 2011 (http://www.mta.info/capconstr/sas/documents/2011_Feb 15_Final.pdf)
[46] "Excavation of West Tunnel for Second Avenue Subway Almost Complete" (http://www.mta.info/news/stories/?story=187). MTAcPress Release. . Retrieved February 5, 2011.
[47] "Boring Manhattan: Ceremony Launches Subway Project" (http://wcbstv.com/local/subway.tunnel.boring.2.1694922.html). *WCBS*. Associated Press. May 14, 2010. . Retrieved May 14, 2010.
[48] Roth, Jamie (May 14, 2010). "Boring for new 2nd Avenue subway begins" (http://abclocal.go.com/wabc/story?section=resources/traffic&id=7441310). *WABC*. . Retrieved May 14, 2010.
[49] "Tunneling for Second Avenue Subway Continues" (http://www.mta.info/news/stories/?story=215). *Metropolitan Transportation Authority*. March 28, 2011. . Retrieved September 24, 2011.
[50] Reproduction of MTA Construction Company schedule sheet (http://1.bp.blogspot.com/_tjgTAe0iBhA/Ssu6TMu7DzI/AAAAAAAACgI/G_oKRm2k9Pg/s1600-h/090924_C1C2AC5ACombined_CB8_final+-+p.3.jpg)
[51] Neuman, William (May 18, 2007). "Manhattan: Budget Increases for New Subway" (http://www.nytimes.com/2007/05/18/nyregion/18mbrfs-subway.html). *The New York Times*. . Retrieved May 24, 2007.
[52] Earle, Geoff (April 18, 2006). "Feds Finally Aboard 2nd Avenue $ubway" (http://www.nypost.com/news/nationalnews/62580.htm). *New York Post*. .
[53] Donohue, Pete (April 24, 2009). "Second Ave. subway set back – again" (http://www.nytimes.com/1988/09/14/nyregion/the-talk-of-fishers-island-isle-s-summer-gentry-squeezing-out-locals.html). *Daily News* (New York). . Retrieved February 20, 2010.

External links

- Second Avenue Subway - Project Home (http://www.mta.info/capconstr/sas/) - the MTA's homepage for this project

nycsubway.org

- The Second Avenue Subway – Historical Reports (http://www.nycsubway.org/lines/2ndave.html)
- Completed Portions of the 2nd Avenue Subway (http://www.nycsubway.org/articles/2ndave-builtfaq.html)
- The Second Avenue Subway Line . . . the line that almost never was (http://www.nycsubway.org/articles/2ndave-neverwas.html)

Other Second Avenue Subway project links

- Second Avenue Subway: A New York Transit Museum on-line exhibit (http://www.transitmuseumeducation.org/fbu/projects/secondavenue)
- *The Launch Box* (http://thelaunchbox.blogspot.com) – A Blog About Building the Second Avenue Subway
- *The Second Avenue Sagas* (http://secondavenuesagas.com/category/second-avenue-subway/) - News from the Second Avenue Subway project

The_Bronx

The Bronx	
— Borough of New York City —	
Bronx County	
 Concourse Village and Yankee Stadium	
Nickname(s): "The Boogie Down Bronx"	
Motto: *"Ne cede malis" - "Yield Not To Evil"*	
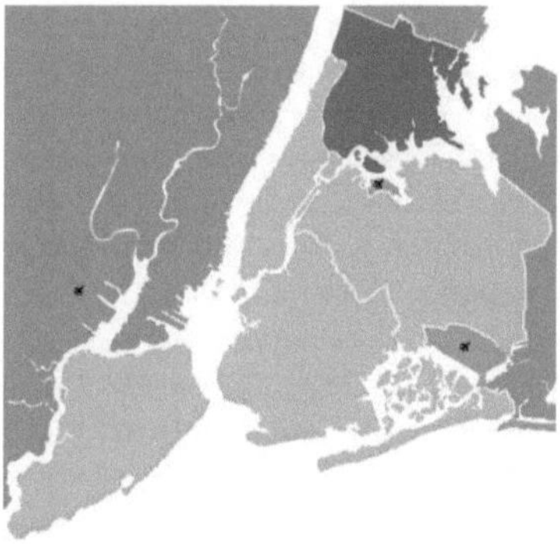 The Bronx is shown in orange.	
Coordinates: 40°50′14″N 73°53′10″W	
Country	United States
State	New York
County	Bronx
City	New York City
Borough created	1898 (**County** in 1914)
Government	

• Type	Borough (New York City)
• Borough President	Ruben Diaz, Jr. — *(Borough of the Bronx)*
• District Attorney	Robert T. Johnson — *(Bronx County)*
Area	
• Total	57 sq mi (150 km^2)
• Land	42 sq mi (110 km^2)
• Water	15 sq mi (40 km^2)
Highest elevation	280 ft (m)
Population (April 1, 2010 U.S. Census)[1]	
• Total	1385108
• Density	32393/sq mi (/km^2)
	(2010 pop. as estimated in March 2011; density is July 2006 est. pop. on land area as of 2000[2])
Time zone	Eastern Standard Time (North America) (UTC-5)
• Summer (DST)	Eastern Daylight Time (UTC-4)
ZIP Code	104 + two digits
Area code(s)	347, 718, 917. 646.
Website	Official website of the Bronx Borough President [3]

The Bronx is the northernmost of the five boroughs of New York City. Coextensive with **Bronx County**, it was the last of the 62 counties of New York State to be incorporated. Located north of Manhattan and Queens, and south of Westchester County, the Bronx is the only borough that is located primarily on the mainland (a very small portion of Manhattan, the Marble Hill neighborhood, is physically located on the mainland, due to the rerouting of the Harlem River in 1897). The Bronx's population is 1,385,108 according to the 2010 United States Census.[1] The borough has a land area of 42 square miles (109 km^2), making it the fourth-largest in land area of the five boroughs, the fourth most populated, and the third-highest in density of population.[2] [4]

The Bronx is divided by the Bronx River into a hillier section in the west, closer to Manhattan, and the flatter East Bronx, closer to Long Island. The West Bronx was annexed to New York City (then largely confined to Manhattan) in 1874, and the areas east of the Bronx River were annexed in 1895.[5] The Bronx first assumed a distinct legal identity when it became a borough of Greater New York in 1898. Bronx County, with the same boundaries as the borough, was separated from New York County (afterwards coextensive with the Borough of Manhattan) as of January 1, 1914.[6]

Although the Bronx is the third most densely populated county in the U.S.,[2] about a quarter of its area is open space,[7] including Woodlawn Cemetery, Van Cortlandt Park, Pelham Bay Park, the New York Botanical Garden and the Bronx Zoo in the borough's north and center, on land deliberately reserved in the late 19th century as urban development progressed northwards and eastwards from Manhattan with the building of roads, bridges and railways.

The Bronx River was named after Jonas Bronck, who created the first settlement as part of the New Netherland colony in 1639,[8] [9] [10] [11] [12] and eventually lent its name to the entire borough. The native Lenape were progressively displaced after 1643 by settlers. The Bronx received many Irish, German, Jewish and Italian immigrants as its once-rural population exploded between the mid-19th and mid-20th centuries. They were succeeded after 1945 by African Americans and Hispanic Americans from the Caribbean basin — especially Puerto

Rico[13] and later the Dominican Republic, Jamaica and Albania. In recent years, this cultural mix has made the Bronx a wellspring of both Latin music and hip hop.

The Bronx contains one of the five poorest Congressional Districts in the U.S., the 16th, but its wide variety of neighborhoods also includes the affluent and middle to upper income Riverdale, Schuylerville and Country Club.[14] [15] The Bronx, particularly the South Bronx, saw a sharp decline in population, livable housing, and the quality of life in the late 1960s and the 1970s, culminating in a wave of arson. Since then the communities have shown significant redevelopment starting in the late 1980s before picking up pace in the 1990s into today.[16]

New York's five boroughs overview				
Jurisdiction		**Population**	**Land area**	
Borough of	*County of*	*1 July 2011 Estimates*	*square miles*	*square km*
Manhattan	New York	1,601,948	23	59
The Bronx	**Bronx**	**1,392,002**	**42**	109
Brooklyn	Kings	2,532,645	71	183
Queens	Queens	2,247,848	109	283
Staten Island	Richmond	470,467	58	151
City of New York		**8,244,910**	**303**	**786**
State of New York		19,465,197	47,214	122,284
Source: United States Census Bureau[1] [2] [17]				

History

The history of the Bronx during the 20th century may be divided into four periods: a boom period during 1900–29, with a population growth by a factor of six from 200,000 in 1900 to 1.3 million in 1930. The Great Depression and post WW2 years saw a slowing of growth leading into an eventual decline. The mid to late century were hard times, as the Bronx declined 1950-85 from a predominantly moderate-income to a predominantly lower-income area with high rates of violent crime and poverty. Finally the Bronx has enjoyed an economic and developmental resurgence starting in the late 1980 that continues into today.[18]

New York City, expands

For generations a mostly rural area of small farms supplying the city markets, the Bronx grew into a railroad suburb in the late 19th century. Faster transportation allowed for rapid population growth in the late 19th century, involving the move from horse-drawn street cars to elevated railways and the subway system, which linked to Manhattan in 1904.[18]

The Grand Concourse & 161st Street at the turn of the 20th century.

The South Bronx was for many years a manufacturing center, and in the early part of the 20th century was noted as a center of piano manufacturing. In 1919, the Bronx was the site of 63 piano factories employing more than 5,000 workers.[19]

At the end of World War I, the Bronx hosted the rather small 1918 World's Fair at 177th Street and DeVoe Avenue.[5] [20]

The Bronx underwent rapid urban growth after World War I. Extensions of the New York City Subway contributed to the increase in population as thousands of immigrants flooded The Bronx, resulting in a major boom in residential construction. Among these groups, many Irish Americans, Italian Americans and especially Jewish Americans settled here. In addition, French, German, Polish and other immigrants moved into the borough. The Jewish population also increased notably during this time. In 1937, according to Jewish organizations, 592,185 Jews lived in The Bronx (43.9% of the borough's population),[21] while only 45,000 Jews lived in the borough in 2002. Many synagogues still stand in the Bronx, but most have been converted to other uses.[22]

Decline

In Prohibition days (1920–33), bootleggers and gangs were active in the Bronx. Irish, Italian and Polish gangs smuggled in most of the illegal whiskey. At this time the oldest sections of the borough began to become poverty stricken areas.

Between 1930 and 1960, moderate and upper income Bronxites (Predominantly non-Hispanic Whites) rapidly began to relocate from the more Southwestern neighborhoods of the borough (White Flight). This migration has left a mostly poor African American and Hispanic (largely Puerto Rican and Dominican) population in the West Bronx. Most predominantly non-Hispanic White communities today are located East and Northwestern sections of the county.

From the mid-1960s to the late-1970s, quality of life for Bronx residents declined sharply. Historians and social scientists have put forward many factors. They include the theory (elaborated in Robert Caro's biography *The Power Broker*)[23] that Robert Moses' Cross-Bronx Expressway destroyed existing residential neighborhoods. Another factor in the Bronx's decline may have been the development of high-rise housing projects.

Yet another may have been a reduction in the real-estate listings and property-related financial services (such as mortgages or insurance policies) offered in some areas of the Bronx — a process known as redlining. Others have suggested a "planned shrinkage" of municipal services, such as fire-fighting.[24] [25] There was also much debate as to whether rent control laws had made it less profitable (or more costly) for landlords to maintain existing buildings with their existing tenants than to abandon or destroy those buildings.[26]

In the 1970s, New York City was also plagued by a wave of arson. The burning of buildings was worst felt in the poorest communities, like the South Bronx. The most common explanation of what occurred was that landlords decided to burn their low property-value buildings and take the insurance money as it was more lucrative to get insurance money than to refurbish or sell a building in a severely distressed area. [27] The Bronx became identified with a high rate of poverty and unemployment, which was mainly a persistent problem in the South Bronx. These Bronx fires and urban decay would paint a face of New York City in the 1970s as a down-and-out city in a state of decline.[28]

Indeed, out of 289 census tracts in the Bronx borough, seven different tracts lost more than 97% of their buildings to fire and abandonment between 1970 and 1980; while another 44 different tracts had more than 50% of their buildings meet the same fate. However, starting in the 1990s, many burned-out and run-down tenements were replaced by multi-unit housing.[28] By the early 1980s, the South Bronx was considered one of the most blighted urban areas in the country, with a loss of 60% of the population and 40% of housing units.

Revitaliztion

Since the late-1980s, significant development has occurred in the Bronx, first stimulated by the city's "Ten-Year Housing Plan"[29] [30] and community members working to rebuild the social, economic and environmental infrastructure by creating affordable housing. Groups affiliated with churches in the South Bronx erected the Nehemiah Homes with about 1,000 units. The grass roots organization Nos Quedamos' endeavor known as Melrose Commons[31] [32] [33] began to rebuild areas in the South Bronx. The ripple effects have been felt borough-wide. The IRT White Plains Road Line began to show an increase in riders. Chains such as Marshalls, Staples, and Target have opened stores in the Bronx. More bank branches have opened in the Bronx as a whole (rising from 106 in 1997 to 149 in 2007), although not primarily in poor or minority neighborhoods, while the Bronx still has fewer branches per person than other boroughs.[34] [35] [36] [37]

In 1997, the Bronx was designated an *All America City* by the National Civic League, acknowledging its comeback from the decline of the mid century. In 2006, *The New York Times* reported that "construction cranes have become the borough's new visual metaphor, replacing the window decals of the 1980s in which pictures of potted plants and drawn curtains were placed in the windows of abandoned buildings."[38] The borough has experienced substantial new building construction since 2002. Between 2002 and June 2007, 33,687 new units of housing were built or were under way and $4.8 billion has been invested in new housing. In the first six months of 2007 alone total investment in new residential development was $965 million and 5,187 residential units were scheduled to be completed. Much of the new development is springing up in formerly vacant lots across the South Bronx.[39]

By 2000, the Bronx had a population of about 1.2 million, and its bridges, highways, and railroads were more heavily traveled than those of any other part of the United States.

Geography

Adjacent counties

- Westchester County – North
- Queens County, New York (Queens) – South
- New York County, New York (Manhattan) – Southwest
- Bergen County, New Jersey – West

Location and physical features

The Bronx is almost entirely situated on the North American mainland.[41] The Hudson River separates the Bronx on the west from Alpine, Tenafly and Englewood Cliffs in Bergen County, New Jersey; the Harlem River separates it from the island of Manhattan to the southwest; the East River separates it from Queens to the southeast; and, to the east, Long Island Sound separates it from Nassau County in western Long Island. Directly north of the Bronx are (from west to east) the adjoining Westchester County communities of Yonkers, Mount Vernon, Pelham Manor and New Rochelle.

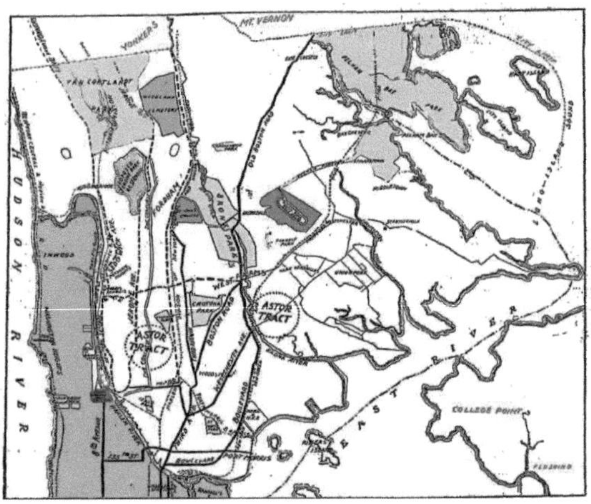

New York Times 1896 map of parks and transit in the newly annexed Bronx. Marble Hill is in pink, cut off by water from the rest of Manhattan in orange. Parks are light green, Woodlawn Cemetery medium green, sports facilities dark green, the not-yet-built Jerome Park Reservoir light blue, St. John's College (now Fordham University) in violet, and the city limits of the newly expanded New York in red.[40]

- (There is also a short southern land boundary with Marble Hill in the Borough of Manhattan, over the filled-in former course of the Spuyten Duyvil Creek. Marble Hill's postal ZIP code, telephonic Area Code and fire service, however, are shared with the Bronx and not Manhattan.)

The New York Public Library maintains a *Map Rectifier* facility that reconciles old maps of the Bronx (and elsewhere) with current cartography.[42]

The Bronx River flows south from Westchester County through the borough, emptying into the East River; it is the only entirely freshwater river in New York City.[43] A smaller river, the Hutchinson River (named after the religious leader Anne Hutchinson, killed along its banks in 1641), passes through the East Bronx and empties into Eastchester Bay.

The Bronx also includes several small islands in the East River and Long Island Sound, such as City Island and Hart Island. Although it is part of the Bronx, Rikers Island in the East River, home to the large jail complex for the entire City, can be reached only by water, by air, or—since 1966—over the Francis Buono Bridge from Queens.

The Bronx's highest elevation 280 feet (85 m), is in the northwest corner, west of Van Cortlandt Park and in the Chapel Farm area near the Riverdale Country School.[44] The opposite (southeastern) side of the Bronx has four large low peninsulas or "necks" of low-lying land that jut into the waters of the East River and were once salt marsh: Hunt's Point, Clason's Point, Screvin's Neck and Throg's Neck. Further up the coastline, Rodman's Neck lies between

Pelham Bay Park in the northeast and City Island.

† (New York City's last freshwater marsh was in Van Cortlandt Park until displaced in the 1930s by the junction of the Mosholu and Henry Hudson Parkways.)

Almost 27%,15.4 square miles (40 km^2) of the Bronx's total area is water, and the irregular shoreline extends for 75 square miles (194 km^2).[45] [46]

Selected Parks and Open Space in the Bronx						
ac- quired	**Name**	acres	square miles	hect- ares	square km	
1863	Woodlawn Cemetery	400	0.6	162	1.6
1888	Pelham Bay Park	2,764	4.3	1,119	11.2
	Van Cortlandt Park	1,146	1.8	464	4.6
	Bronx Park	718	1.1	291	2.9
	Crotona Park	128	0.2	52	0.5
1890	Jerome Park Reservoir	94	0.15	38	0.4
1897	St. James Park	11	0.02	4.6	0.0
1899	Macomb's Dam Park †	28	0.04	12	0.1
1909	Henry Hudson Park	9	0.01	4	0.0
1937	Ferry Point Park	414	0.65	168	1.7
	Soundview Park	196	0.31	79	0.8
1962	Wave Hill	21	0.03	8.5	0.1
Land area of the Bronx in 2000		*26,897*	*42.0*	*10,885*	*108.8*
Water area		*9,855*	*15.4*	*3,988*	*39.9*
Total area [45]		*36,752*	*57.4*	*14,873*	*148.7*
† *closed in 2007 to build a new park & Yankee Stadium* [47]					
Main source: New York City Department of Parks & Recreation [48]					

Parks and open space

Although, in 2006, it was the third most densely populated county in the United States (after Manhattan and Brooklyn),[2] about one-fifth of the Bronx's area, and one-quarter of its land area, is given over to park land: about 7000 acres (km^2).[7]

Woodlawn Cemetery, one of the largest cemeteries in New York City, sits on the western bank of the Bronx River near Yonkers. It opened in 1863, at a time when the Bronx was still considered a rural area.

The northern side of the borough includes the largest park in New York City - Pelham Bay Park, which includes Orchard Beach - and the fourth largest, Van Cortlandt Park, which is west of Woodlawn Cemetery and borders Yonkers.

Nearer the borough's center, and along the Bronx River, is Bronx Park. Its northern end houses the New York Botanical Gardens, which preserve the last patch of the original hemlock forest that once covered the entire city, and its southern end the Bronx Zoo, the largest urban zoological gardens in the U.S.[49]

Farther south is Crotona Park, home to a 3.3-acre (ha) lake, 28 species of trees and a large swimming pool.[50] The land for these parks, and many others, was bought by New York City in 1888, while land was still open and inexpensive, in anticipation of future needs and future pressures for development.[51]

Some of the acquired land was set aside for the Grand Concourse and Pelham Parkway, the first of a series of boulevards and parkways (thoroughfares lined with trees, vegetation and greenery). Later projects included the Bronx River Parkway, which developed a road while restoring the riverbank and reducing pollution, Mosholu Parkway and the Henry Hudson Parkway.

Just south of Van Cortlandt Park is the Jerome Park Reservoir, surrounded by 2 miles (3 km) of stone walls and bordering several small parks in the Bedford Park neighborhood. The reservoir was built in the 1890s on the site of the former Jerome Park Racetrack.[52]

The Northern tip of Hunter Island in Pelham Bay Park.

In 2006, a five-year, $220-million program of capital improvements and natural restoration in 70 Bronx parks was begun (financed by water and sewer revenues) as part of an agreement that allowed a water filtration plant under Van Cortlandt Park's golf course. One major focus is on opening more of the Bronx River's banks and restoring them to a natural state.[53]

Wave Hill, the former estate of George W. Perkins — known for a historic house, gardens, changing site-specific art installations and concerts — overlooks the New Jersey Palisades from a promontory on the Hudson in Riverdale.

Neighborhoods

The number, locations and boundaries of the Bronx's neighborhoods (many of them sitting on the sites of 19th-century villages) have become unclear with time and successive waves of newcomers. In 2006, Manny Fernandez of *The New York Times* wrote,

> "According to a Department of City Planning map of the city's neighborhoods, the Bronx has 49. The map publisher Hagstrom identifies 69. The borough president, Adolfo Carrión Jr., says 61. The Mayor's Community Assistance Unit, in a listing of the borough's community boards, names 68. Wikipedia, the online encyclopedia, lists 44."[54]

Notable Bronx neighborhoods include the South Bronx, Little Italy on Arthur Avenue in the Belmont section, and Riverdale.

East Bronx

(Bronx Community Boards 9 *[south central]*, 10 *[east]*, 11 *[east central]* and 12 *[north central]*)[55]

East of the Bronx River, the borough is relatively flat, and includes four large low peninsulas, or 'necks,' of low-lying land which jut into the waters of the East River and were once saltmarsh: Hunts Point, Clason's Point, Screvin's Neck (Castle Hill Point) and Throgs Neck. The East Bronx has older tenement buildings, low income public housing complexes, and multifamily homes, as well as smaller and larger single family homes. It includes New York City's largest park: Pelham Bay Park along the Westchester-Bronx border.

The neighborhood of Co-op City is the largest cooperative housing development in the world.

Neighborhoods include: Clason's Point, Harding Park, Soundview, Castle Hill, Parkchester *(under Board 9)*, Throggs Neck, Country Club, City Island, Pelham Bay, Co-op City *(Board 10)*, Westchester Square, Van Nest,

Pelham Parkway, Morris Park *(Board 11)*, Williamsbridge, Eastchester, Baychester, Edenwald and Wakefield *(Board 12)*.

City Island and Hart Island

(Bronx Community Board 10)

City Island is located east of Pelham Bay Park in Long Island Sound, and is known for its seafood restaurants and waterfront private homes. City Island's single shopping street, City Island Avenue, is reminiscent of a small New England town. It is connected to Rodman's Neck on the mainland by the City Island Bridge.

East of City Island is Hart Island which is uninhabited and not open to the public. It once served as a prison and now houses New York City's Potter's Field or pauper's graveyard for unclaimed bodies.

The sea is a part of everyday life in City Island.

West Bronx

(Bronx Community Boards 1 to 8, progressing roughly from south to northwest)

The western parts of the Bronx are hillier and are dominated by a series of parallel ridges, running south to north. The West Bronx has older apartment buildings, low income public housing complexes, multifamily homes in its lower income areas as well as larger single family homes in more affluent areas such as Riverdale and Fieldston.[56] It includes New York City's fourth largest park: Van Cortlandt Park along the Westchester-Bronx border. The Grand Concourse, a wide boulevard, runs through it, north to south.

The Grand Concourse at East 165th Street.

Northwestern Bronx

(Bronx Community Boards 7 *[between the Bronx and Harlem Rivers]* and 8 *[facing the Hudson River]* — plus part of Board 12)

Neighborhoods include: Fordham-Bedford, Bedford Park, Norwood, Kingsbridge Heights *(Board 7)*, Kingsbridge, Riverdale *(Board 8)*, and Woodlawn *(Board 12)*. (Marble Hill, Manhattan is now connected by land to the Bronx rather than Manhattan and is served by Bronx Community Board 8.)

South Bronx

(Bronx Community Boards 1 to 6 plus part of Board 7 —— *progressing northwards, Boards 2, 3 and 6 border the Bronx River from its mouth to Bronx Park, while 1, 4, 5 and 7 face Manhattan across the Harlem River*)

Yankee Stadium is located on 161st and River Avenue.

Like other neighborhoods in New York City, the South Bronx has no official boundaries. The name has been used to represent poverty in the Bronx and applied to progressively more northern places so that by the 2000s Fordham Road was often used as a northern limit. The Bronx River more consistently forms an eastern boundary. The South Bronx has many high-density apartment buildings, low income public housing complexes, and multi-unit homes. The South Bronx is home to the Bronx County Courthouse, Borough Hall, and other government buildings, as well as Yankee Stadium. The Cross Bronx Expressway bisects it, east to west. The South Bronx has some of the poorest neighborhoods in the country, as well as very high crime areas.

Neighborhoods include: The Hub (a retail district at Third Avenue and East 149th Street), Port Morris, Mott Haven *(Board 1)*, Melrose *(Board 1 & Board 3)*, Morrisania, East Morrisania [also known as Crotona Park East] *(Board 3)*, Hunts Point, Longwood *(Board 2)*, Highbridge, Concourse *(Board 4)*, West Farms, Belmont, East Tremont *(Board 6)*, Tremont, Morris Heights *(Board 5)*, University Heights, and Fordham *(Board 5 & Board 7)*.

Shopping districts

The Hub on Third Avenue.

Prominent shopping areas in the Bronx include Fordham Road, Bay Plaza (in Co-op City), The Hub, Riverdale/Kingsbridge Shopping center and Bruckner Boulevard. Shops are also concentrated on streets aligned underneath elevated railroad lines, including Westchester Avenue, White Plains Road, Jerome Avenue, Southern Boulevard and Broadway. The Gateway Center at Bronx Terminal Market contains several big-box stores, which opened in 2009 south of Yankee Stadium.

The Bronx Hub

The Hub–Third Avenue Business Improvement District (B.I.D.) is the retail heart of the South Bronx, located where four roads converge: East 149th Street, Willis, Melrose and Third Avenues.[57] It is primarily located inside the neighborhood of Melrose but also lines the northern border of Mott Haven.[58] The Hub has been called "the Broadway of the Bronx."[59] It is the site of both maximum traffic and architectural density. In configuration, it resembles a miniature Times Square, a spatial "bow-tie" created by the geometry of the street.[60] The area is part of Bronx Community Board 1.

The Gateway Center at Bronx Terminal Market

The Gateway Center at Bronx Terminal Market is a shopping center in The Bronx that encompasses less than one million square feet of retail space, built on a 17-acre (69,000 m2) site that formerly held the Bronx Terminal Market, a wholesale fruit and vegetable market as well as the former Bronx House of Detention, south of Yankee Stadium.[61]

The $500 million shopping center, which was completed in 2009, saw the construction of new buildings and two smaller buildings, one new and the other a renovation of an existing building that was part of the original market. The two main buildings are linked by a six-level garage for 2,600 cars. The center has earned itself a LEED "Silver" designation in its design.[62]

An aerial view of the Bronx, Harlem River, Harlem, Hudson River, and George Washington Bridge.

Street scene on Fordham Road, a major street in The Bronx.

Transportation

Roads and streets

Bronx-Whitestone Bridge

The Bronx street grid is irregular. Like the northernmost part of upper Manhattan, the West Bronx's hilly terrain leaves a relatively free-style street grid. Much of the West Bronx's street numbering carries over from upper Manhattan, but does not match it exactly; East 132nd Street is the lowest numbered street in the Bronx. This dates from the mid-19th century when the southwestern area of Westchester County west of the Bronx River, was incorporated into New York City and known as the Northside.

The East Bronx is considerably flatter, and the street layout tends to be more regular. Only the Wakefield neighborhood picks up the street numbering, albeit at a disalignment due to Tremont Avenue's layout. At the same diagonal latitude, West 262nd Street in Riverdale matches East 237th Street in Wakefield.

Three major north-south thoroughfares run between Manhattan and the Bronx: Third Avenue, Park Avenue, and Broadway. Other major north-south roads include the Grand Concourse, Jerome Avenue, Sedgwick Avenue, Webster Avenue, and White Plains Road. Major east-west thoroughfares include Mosholu Parkway, Gun Hill Road, Fordham Road, Pelham Parkway, and Tremont Avenue.

Most east-west streets are prefixed with either *East* or *West*, to indicate on which side of Jerome Avenue they lie (continuing the similar system in Manhattan, which uses Fifth Avenue as the dividing line).

The historic Boston Post Road, part of the long pre-revolutionary road connecting Boston with other northeastern cities, runs east-west in some places, and sometimes northeast-southwest.

Mosholu and Pelham Parkways, with Bronx Park between them, Van Cortlandt Park to the west and Pelham Bay Park to the east, are also linked by bridle paths.

Approximately 61.6% of all Bronx households do not have access to a car. Citywide, the percentage of autoless households is 55%.[63]

Highways

Several major limited access highways traverse the Bronx. These include:

- the Bronx River Parkway
- the Bruckner Expressway (I-278/I-95)
- the Cross-Bronx Expressway (I-95/I-295)
- the New England Thruway (I-95)
- the Henry Hudson Parkway (NY-9A)
- the Hutchinson River Parkway
- the Major Deegan Expressway (New York Thruway) (I-87)

Bridges and tunnels

Many bridges and tunnels connect the Bronx to Manhattan and Queens (3). These include, from west to east:

An aerial View of the Throgs Neck Bridge.

To Manhattan: the Spuyten Duyvil Bridge, the Henry Hudson Bridge, the Broadway Bridge, the University Heights Bridge, the Washington Bridge, the Alexander Hamilton Bridge, the High Bridge, the Concourse Tunnel, the Macombs Dam Bridge, the 145th Street Bridge, the 149th Street Tunnel, the Madison Avenue Bridge, the Park Avenue Bridge, the Lexington Avenue Tunnel, the Third Avenue Bridge (southbound traffic only), and the Willis Avenue Bridge (northbound traffic only).

To Manhattan or Queens: the Triborough Bridge, officially known as the Robert F. Kennedy Bridge.

To Queens: the Bronx Whitestone Bridge and the Throgs Neck Bridge.

Mass transit

Middletown Road subway station on the 6 <6> trains.

The Bronx is served by six lines of the New York City Subway with 70 stations in the Bronx:

- IND Concourse Line (B D trains)
- IRT Broadway – Seventh Avenue Line (1 train)
- IRT Dyre Avenue Line (5 train)
- IRT Jerome Avenue Line (4 train)
- IRT Pelham Line (6 <6> trains)
- IRT White Plains Road Line (2 5 trains)

Two Metro-North Railroad commuter rail lines (the Harlem Line and the Hudson Line) serve 11 stations in the Bronx. (Marble Hill, between the Spuyten Duyvil and University Heights stations, is actually in the only part of Manhattan connected to the mainland.) In addition, trains serving the New Haven Line stop at Fordham Road.

Postal service

The United States Postal Service operates post offices in the Bronx. The Bronx General Post Office is located at 558 Grand Concourse.[64] [65]

NYC Transit bus operating on the Bx40 line in University Heights.

Demographics

Race, ethnicity, language, and immigration

According to the 2010 Census, 53.5% of Bronx's population was of Hispanic, Latino, or Spanish origin (they may be of any race); 30.1% non-Hispanic Black or African American, 10.9% of the population was non-Hispanic White, 3.4% non-Hispanic Asian, 0.6% from some other race (non-Hispanic) and 1.2% of two or more races (non-Hispanic). The U.S. Census considers the Bronx to be the most diverse area in the country. There is an 89.7 percent chance that any two residents, chosen at random, would be of different race or ethnicity.[66] The borough's formerly most populous ethnic group, white, declined from 98.3% in 1940 to 35.7% by 1990.[67]

The Bronx is the only New York City borough with a Hispanic majority, many of whom are Puerto Ricans and Dominicans. In 2000, The Bronx had some of the nation's highest percentages of Puerto Ricans and Dominicans with 24.0% and 10.0%, respectively.[68]

According to the 2009 American Community Survey, White Americans of both Hispanic and non-Hispanic origin represented over one-fifth (22.9%) of The Bronx's population. However, non-Hispanic whites formed under one-eighth (12.1%) of the population, down from 34.4% in 1980.[67] Out of all five boroughs, The Bronx has the lowest number and percentage of white residents. 320,640 whites called the Bronx home, of which 168,570 were non-Hispanic whites. The majority of the non-Hispanic European American population is of Italian and Irish descent. People of Italian descent numbered over 55,000 individuals and made up 3.9% of the population. People of Irish descent numbered over 43,500 individuals and made up 3.1% of the population. German Americans and Polish Americans made up 1.4% and 0.8% of the population respectively.

At the 2009 American Community Survey, Black Americans made the second largest group in the Bronx after Hispanics and Latinos. Blacks of both Hispanic and non-Hispanic origin represented over one-third (35.4%) of the Bronx's population. Blacks of non-Hispanic origin made up 30.8% of the population. Over 495,200 blacks resided in the borough, of which 430,600 were non-Hispanic blacks. Over 61,000 people identified themselves as "Sub-Saharan African" in the survey, making up 4.4% of the population.

Native Americans are a very small minority in the borough. Only some 5,560 individuals (out of the borough's 1.4 million people) are Native American, which is equal to just 0.4% of the population. In addition, roughly 2,500 people are Native Americans of non-Hispanic origin.

Historical populations			
Census	**Pop.**		**%±**
1790	1781		—
1800	1755		−1.5%
1810	2267		29.2%
1820	2782		22.7%
1830	3023		8.7%
1840	5346		76.8%
1850	8032		50.2%
1860	23593		193.7%
1870	37393		58.5%
1880	51980		39.0%
1890	88908		71.0%
1900	200507		125.5%
1910	430980		114.9%
1920	732016		69.8%
1930	1265258		72.8%
1940	1394711		10.2%
1950	1451277		4.1%
1960	1424815		−1.8%
1970	1471701		3.3%
1980	1168972		−20.6%
1990	1203789		3.0%
2000	1332650		10.7%
2010	1385108		3.9%
Est. 2011	1392002	[69]	0.5%
Sources: 1790–1990;[70] 2000–2010[1]			

In 2009, Hispanic and Latino Americans represented 52.0% of the Bronx's population. Puerto Ricans represented 23.2% of the borough's population. Over 72,500 Mexicans lived in the Bronx, and they formed 5.2% of the population. Cubans numbered over 9,640 members and formed 0.7% of the population. In addition, over 319,000 people were of various Hispanic and Latino groups, such as Dominican, Salvadoran, and so on. These groups collectively represented 22.9% of the population. At the 2010 Census, 53.5% of Bronx's population was of Hispanic, Latino, or Spanish origin (they may be of any race). Asian Americans are a small but sizable minority in the borough. Roughly 49,600 Asians make up 3.6% of the population. Roughly 13,600 Indians call The Bronx home, along with 9,800 Chinese, 6,540 Filipinos, 2,260 Vietnamese, 2,010 Koreans, and 1,100 Japanese.

Multiracial Americans are also a sizable minority in the Bronx. People of multiracial heritage number over 41,800 individuals and represent 3.0% of the population. People of mixed Caucasian and African American heritage number over 6,850 members and form 0.5% of the population. People of mixed Caucasian and Native American heritage number over 2,450 members and form 0.2% of the population. People of mixed Caucasian and Asian heritage number over 880 members and form 0.1% of the population. People of mixed African American and Native American heritage number over 1,220 members and form 0.1% of the population.

Approximately 44.3% of the population over the age of 5 speak only English at home, which is roughly 570,000 people. The majority (55.7%) of the population speak non-English languages at home. Over 580,600 people (45.2% of the population) speak Spanish at home.[71] [72]

According to the 2005–2007 American Community Survey Estimates, the borough's population was 23.0% White (13.0% non-Hispanic White alone), 34.5% Black or African American (30.6% non-Hispanic Black or African American alone), 0.7% American Indian and Alaska Native, 3.8% Asian, 0.1% Native Hawaiian and Other Pacific Islander, 40.4% from some other race and 2.4% from two or more races. 50.7% of the total population were Hispanic or Latino of any race (23.3% of Bronx's population were Puerto Ricans).[73] 31.7% of the population were foreign born and another 8.9% were born in Puerto Rico, U.S. Island areas, or born abroad to American parents. 55.6% spoke a language other than English at home and 16.4% had a Bachelor's degree or higher.[74]

The Census of 1930 counted only 1.0% (12,930) of the Bronx's population as Negro (while making no distinct counts of Hispanic or Spanish-surname residents).[75]

Immigrants from Ghana have clustered along the Grand Concourse.[76]

Based on sample data from the 2000 census, the U.S. Census Bureau estimated that 47.3% of the population five and older spoke only English at home, while 43.7% spoke Spanish at home, either exclusively or along with English. Other languages or groups of languages spoken at home by more than 0.25% of the population of the Bronx include Italian (1.36%), Kru, Igbo, or Yoruba [West Africa] (0.72%) and French (0.54%).

Bronx residents born abroad or overseas, 1930 and 2000					
1930 United States Census[75]			**2000 United States Census[77]**		
Total population of the Bronx	**1,265,258**		**Total population of the Bronx**	**1,332,650**	
			All born abroad or overseas ‡	**524,410**	***39.4%***
			Puerto Rico	126,649	*9.5%*
Foreign-born Whites	**477,342**	***37.7%***	***All foreign-born***	**385,827**	***29.0%***
White persons born in Russia	135,210	*10.7%*	Dominican Republic	124,032	*9.3%*
White persons born in Italy	67,732	*5.4%*	Jamaica	51,120	*3.8%*
White persons born in Poland	55,969	*4.4%*	Mexico	20,962	*1.6%*
White persons born in Germany	43,349	*3.4%*	Guyana	14,868	*1.1%*
White persons born in the Irish Free State †	34,538	*2.7%*	Ecuador	14,800	*1.1%*
Other foreign birthplaces of Whites	140,544	*11.1%*	Other foreign birthplaces	160,045	*12.0%*
† the 26 counties now within the Republic of Ireland			‡ beyond the 50 states & District of Columbia		

Population and housing

At the 2010 Census, there were, 1,385,108 people living in Bronx, a 3.9% increase since 2000. As of the United States Census[78] of 2000, there were 1,332,650 people, 463,212 households, and 314,984 families residing in the borough. The population density was 31,709.3 inhabitants per square mile (12,242.2/km²). There were 490,659 housing units at an average density of 11,674.8 per square mile (4,507.4/km²).[78]

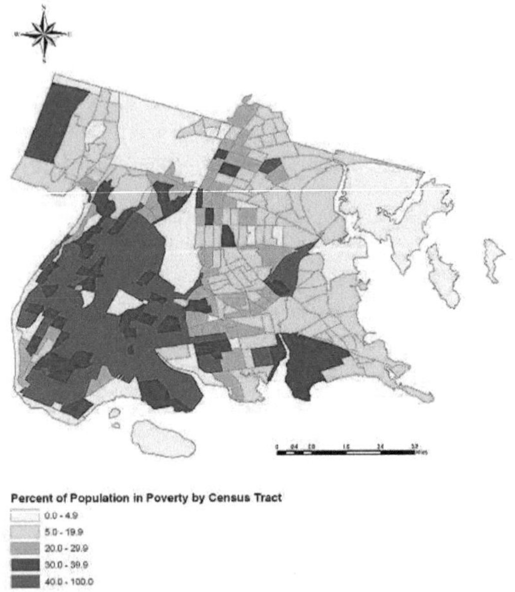

Poverty concentrations within the Bronx, by Census Tract.

There were 463,212 households out of which 38.1% had children under the age of 18 living with them, 31.4% were married couples living together, 30.4% had a female householder with no husband present, and 32.0% were non-families. 27.4% of all households were made up of individuals and 9.4% had someone living alone who was 65 years of age or older. The average household size was 2.78 and the average family size was 3.37.[78]

The age distribution of the population in the Bronx was as follows: 29.8% under the age of 18, 10.6% from 18 to 24, 30.7% from 25 to 44, 18.8% from 45 to 64, and 10.1% 65 years of age or older. The median age was 31 years. For every 100 females there were 87.0 males.[78]

Individual and household income

The 1999 median income for a household in the borough was $27,611, and the median income for a family was $30,682. Males had a median income of $31,178 versus $29,429 for females. The per capita income for the borough was $13,959. About 28.0% of families and 30.7% of the population were below the poverty line, including 41.5% of those under age 18 and 21.3% of those age 65 or over.

Government and politics

Local government

Since New York City's consolidation in 1898, the Bronx has been governed by the New York City Charter that provides for a mayor-council system. The centralized New York City government is responsible for all municipal functions.

The office of Borough President was created in the consolidation of 1898 with powers mostly derived from having a vote on the New York City Board of Estimate. The 1989 Board of Estimate of City of New York v. Morris case declared the Board unconstitutional, and since 1990 the Borough President has acted as an advocate for the borough at the mayoral agencies, the City Council, the New York state government, and corporations.

On February 18, 2009, U.S. President Barack Obama appointed the former Bronx Borough President, Adolfo Carrión, Jr., to the position of Director of the White House Office of Urban Affairs Policy.[79]

On April 21, 2009, a special election was held to choose Carrión's successor. Democratic New York State Assembly member Rubén Díaz, Jr., running on the "Bronx Unity" ticket, won this election with 29,420 votes (86%) against 4,646 votes (14%) for the Republican Anthony Ribustello ("People First") and 11 votes for write-in candidates.[80] [81] [82] On May 1, 2009, Assemblyman Diaz was sworn in as the 13th Borough President of the Bronx (For his predecessors, see the List of Bronx Borough Presidents.)

Every currently elected public official in the Bronx has first won the Democratic nomination (whether or not also nominated by other parties). Local party platforms center on affordable housing, education and economic development. Controversial political issues in the Bronx include environmental issues, the cost of housing, and the alienation of parkland for new Yankee Stadium.

Since its separation from New York County on January 1, 1914, the Bronx, has had, like each of the other 61 counties of New York State, its own directly elected District Attorney, the county's chief public prosecutor.[6] Robert T. Johnson, a Democrat, has been the District Attorney of Bronx County since 1989. He was the first African-American District Attorney in New York State.[83]

Eight members of the New York City Council represent districts wholly within the Bronx, while a ninth represents a Manhattan district (8) that also includes a small area of the Bronx. (All of them were Democrats in 2008.) One of those members, Joel Rivera (District 15), has been the Council's Majority Leader since 2002.

The Bronx also has 12 Community Boards, appointed bodies that field complaints and advise on land use and municipal facilities and services for local residents, businesses and institutions. (They are listed at Bronx Community Boards).

Legislative and congressional representatives

In 2008, three Democrats represented almost all of the Bronx in the United States House of Representatives.

- José M. Serrano (first elected in March 1990) represents New York's 16th congressional district, which covers much of the South Bronx. It was, in 2000, the poorest of the nation's 435 districts (42.8% below the poverty line); it was also the most Hispanic of New York state's 29 congressional districts (62.8%) and the youngest (34.5% under 18 years old; 6.7% over 65).
- Eliot Engel (first elected in 1988) represents the 17th District which includes parts of the northwest Bronx as well as parts of Westchester and Rockland counties.
- Joseph Crowley (first elected in 1998) represents the 7th District which spans the East Bronx and includes Co-op City, City Island, Pelham Bay, Morris Park, Pelham Parkway, Parkchester, Castle Hill and Throgs Neck, as well as parts of northwest Queens.
- *(Riker's Island, the city's main jail complex, is included in the 15th District, which covers Upper Manhattan and utilities facilities in Astoria, Queens. It is represented by Charles B. Rangel, first elected in 1970. In 2006, the Congressional election returns in this district included no votes from the Bronx or Queens.)*

All of these Representatives won over 75% of their districts' respective votes in both 2004 and 2006. *National Journal's* neutral rating system placed all of their voting records in 2005 and 2006 somewhere between very liberal and extremely liberal.[14] [15]

11 out of 150 members of the New York State Assembly (the lower house of the state legislature) represent districts wholly within the Bronx. Six State Senators out of 62 represent Bronx districts, half of them wholly within the County, and half straddling other counties. All these legislators are Democrats who won between 65% and 100% of their districts' vote in 2006.[84]

Votes for other offices

In the 2004 presidential election, Senator John F. Kerry (Democratic/Working Families) received 81.8% of the vote in the Bronx while President George W. Bush (Republican/Conservative) received 16.3%.

A year later, the Democratic former Bronx Borough President Fernando Ferrer won 59.8% of the borough's vote against 38.8% for Mayor Michael Bloomberg (Republican/Independence) who carried every other borough in his winning campaign for re-election. In 2009, the Bronx voted slightly more strongly against the successful re-election of Mayor Bloomberg, who received 37% of the Bronx vote (as an independent supported by the Republican and Independence Parties) against 61.2% for New York City Comptroller Bill Thompson (Democratic and Working Families).

In 2006, successfully reelected Senator Hillary Clinton (Democratic, Working Families & Independence) won 89.5% of the Bronx's vote against 9.6% for Yonkers ex-Mayor John Spencer (Republican and Conservative), while Eliot Spitzer (Democratic, Working Families & Independence) received 88.8% of the Borough's vote in winning the Governorship against John Faso (Republican & Conservative), who received 9.7% of the Bronx's vote.

In the Presidential primary elections of February 5, 2008, Sen. Clinton won 61.2% of the Bronx's 148,636 Democratic votes against 37.8% for Barack Obama and 1.0% for the other four candidates combined. At the same time, John McCain won 54.4% of the borough's 5,643 Republican votes, Mitt Romney 20.8%, Mike Huckabee 8.2%, Ron Paul 7.4%, Rudy Giuliani 5.6%, and the other three candidates 3.6% between them.[85]

In the Presidential general election of November 4, 2008, Sen. Obama won 87.8% of the Bronx's vote (338,261) on the Democratic and Working Families Party lines of the ballot, Sen. McCain won 10.8% (41,683) on the Republican, Independence and Conservative Party lines, and other candidates won 1.3% (1,342) between them. The Democratic candidate's percentage of the Presidential vote increased by 6% from 2004, while the Republican presidential candidate's percentage declined by 5.5%.

After becoming a separate county in 1914, the Bronx has supported only two Republican Presidential candidates. It voted heavily for the winning Republican Warren G. Harding in 1920, but much more narrowly on a split vote for his victorious Republican successor Calvin Coolidge in 1924 (Coolidge 79,562; John W. Davis, Dem., 72,834; Robert La Follette, 62,202 equally divided between the Progressive and Socialist lines).

Since then, the Bronx has always supported the Democratic Party's nominee for President, starting with a vote of 2–1 for the unsuccessful Al Smith in 1928, followed by four 2-1 votes for the successful Franklin D. Roosevelt. (Both had been Governors of New York, but the Bronx voted against two former Republican Governors who ran for President: Charles Evans Hughes in 1916 and Thomas E. Dewey in 1944 and 1948.)[86]

The Bronx has often shown striking differences from other boroughs in elections for Mayor. The only Republican to carry the Bronx since 1914 was Fiorello La Guardia in 1933, 1937 and 1941 (and in the latter two elections, only because his 30–32% vote on the American Labor Party line was added to 22–23% as a Republican).[87] The Bronx was thus the only borough not carried by the successful Republican re-election campaigns of Mayors Rudolph Giuliani in 1997 and Michael Bloomberg in 2005. The anti-war Socialist campaign of Morris Hillquit in the 1917 mayoral election won over 31% of the Bronx's vote, putting him second and well ahead of the 20% won by the incumbent pro-war Fusion Mayor John P. Mitchel, who outpolled Hillquit city-wide by 23.2% to 21.7%.[88]

Education

Education in the Bronx is provided by a large number of public and private institutions, many of which draw students who live beyond the Bronx. The New York City Department of Education manages public noncharter schools in the borough. In 2000, public schools enrolled nearly 280,000 of the Bronx's residents over 3 years old (out of 333,100 enrolled in all pre-college schools).[89] There are also several public charter schools. Private schools range from élite independent schools to religiously affiliated schools run by the Roman Catholic Archdiocese of New York and Jewish organizations.

Educational attainment: In 2000, according to the U.S. Census, out of the nearly 800,000 people in the Bronx who were then at least 25 years old, 62.3% had graduated from high school and 14.6% held a bachelor's or higher college degree. These percentages were lower than those for New York's other boroughs, which ranged from 68.8% (Brooklyn) to 82.6% (Staten Island) for high school graduates over 24, and from 21.8% (Brooklyn) to 49.4% (Manhattan) for college graduates. (The respective state and national percentages were *[NY]* 79.1% & 27.4% and *[US]* 80.4% & 24.4%.)[90]

High schools

In the 2000 Census, 79,240 of the nearly 95,000 Bronx residents enrolled in high school attended public schools.[89]

The Bronx High School of Science

Many public high schools are located in the borough including the élite Bronx High School of Science, DeWitt Clinton High School, High School for Violin and Dance, Bronx Leadership Academy 2, Bronx International High School, the School for Excellence, the Morris Academy for Collaborative Study, Wings Academy for young adults, The Bronx School for Law, Government and Justice, Validus Preparatory Academy, Bronx Expeditionary Learning High School, Bronx Academy of Letters, Herbert H. Lehman High School and High School of American Studies. The Bronx is also home to three of New York City's most prestigious private, secular schools: Fieldston, Horace Mann, and Riverdale Country School.

High schools linked to the Roman Catholic Church include: Saint Raymond's Academy for Girls, All Hallows High School, Fordham Preparatory School, Monsignor Scanlan High School, St. Raymond High School for Boys, Cardinal Hayes High School, Cardinal Spellman High School, The Academy of Mount Saint Ursula, Aquinas High School, Preston High School, St. Catharine Academy, Mount Saint Michael Academy, and St. Barnabas High School.

The SAR Academy and SAR High School are Modern Orthodox Jewish Yeshiva coeducational day schools in Riverdale, with roots in Manhattan's Lower East Side.

In the 1990s, New York City began closing the large, public high schools in the Bronx and replacing them with small high schools. Among the reasons cited for the changes were poor graduation rates and concerns about safety. Schools that have been closed or reduced in size include John F. Kennedy, James Monroe, Taft, Theodore Roosevelt, Adlai Stevenson, Evander Childs, Christopher Columbus, Morris, Walton, and South Bronx High Schools. More recently the City has started phasing out large middle schools, also replacing them with smaller schools.

Institutions of higher education

Fordham University's Keating Hall.

In 2000, 49,442 (57.5%) of the 86,014 Bronx residents seeking college, graduate or professional degrees attended public institutions.[89]

Several colleges and universities are located in the Bronx.

Fordham University, was founded as St. John's College in 1841 by the Diocese of New York as the first Catholic institution of higher education in the northeast. It is now officially an independent institution, but strongly embraces its Jesuit heritage. The 85-acre (m^2) Bronx campus, known as Rose Hill, is the main campus of the university, and is among the largest within the city (other Fordham campuses are located in Manhattan and Westchester County).[49]

Three campuses of the City University of New York are in the Bronx: Hostos Community College, Bronx Community College (occupying the former University Heights Campus of New York University)[91] and Herbert H. Lehman College (formerly the uptown campus of Hunter College), which offers both undergraduate and graduate degrees.

The College of Mount Saint Vincent is a Catholic liberal arts college in Riverdale under the direction of the Sisters of Charity of New York. Founded in 1847 as a school for girls, the academy became a degree-granting college in 1911 and began admitting men in 1974. The school serves 1,600 students. Its campus is also home to the Academy for Jewish Religion, a transdenominational rabbinical and cantorial school.

Manhattan College is a Catholic college in Riverdale which offers undergraduate programs in the arts, business, education, engineering, and science. It also offers graduate programs in education and engineering.

Albert Einstein College of Medicine, part of Yeshiva University, is in Morris Park.

Two colleges based in Westchester County have Bronx campuses. The Catholic and nearly-all-female College of New Rochelle maintains satellite campuses at Co-op City and in The Hub. The coeducational and non-sectarian Mercy College in Dobbs Ferry, founded by the Catholic Sisters of Mercy in 1950, has a campus near Westchester Square.

By contrast, the private, proprietary Monroe College, focused on preparation for business and the professions, started in the Bronx in 1933 but now has a campus in New Rochelle (Westchester County) as well the Bronx's Fordham neighborhood.[92]

The State University of New York Maritime College in Fort Schuyler (Throggs Neck)—at the far southeastern tip of the Bronx—is the national leader in maritime education and houses the Maritime Industry Museum. (Directly across Long Island Sound is Kings Point, Long Island, home of the United States Merchant Marine Academy and the American Merchant Marine Museum.)

Cultural life and institutions

Author Edgar Allan Poe spent the last years of his life (1846 to 1849) in the Bronx at Poe Cottage, now located at Kingsbridge Road and the Grand Concourse. A small wooden farmhouse built around 1812, the cottage once commanded unobstructed vistas over the rolling Bronx hills to the shores of Long Island.[94]

The Bronx's P.L.A.Y.E.R.S. Club Steppers performing at the 2007 Fort Greene Park Summer Literary Festival in Brooklyn. (Note the T-shirts' inscription "I ♥ BX" *[Bronx]*, echoing the ubiquitous slogan "I ♥ NY" *[I Love New York]*).[93]

The Bronx's evolution from a hot bed of Latin jazz to an incubator of hip hop was the subject of an award-winning documentary, produced by City Lore and broadcast on PBS in 2006, "From Mambo to Hip Hop: A South Bronx Tale" [95]. Hip Hop first emerged in the South Bronx in the early 1970s. *The New York Times* has identified 1520 Sedgwick Avenue "an otherwise unremarkable high-rise just north of the Cross Bronx Expressway and hard along the Major Deegan Expressway" as a starting point, where DJ Kool Herc presided over parties in the community room.[96] [97]

On August 11, 1973, DJ Kool Herc was a D.J. and M.C. at a party in the recreation room of 1520 Sedgewick Avenue in the Bronx adjacent to the Cross-Bronx Expressway.[98] While it was not the actual "Birthplace of Hip Hop" – the genre developed slowly in several places in the 1970s – it was verified to be the place where *one of* the pivotal and formative events occurred.[98] Specifically, DJ Kool Herc:

> extended an instrumental beat (mixing or scratching) to let people dance longer (B-boying) and began MC'ing (rapping) during the extended breakdancing. ... [This] helped lay the foundation for a cultural revolution.
>
> —History Detectives[98]

Beginning with the advent of beat match DJ'ing, in which Bronx DJs (Disc Jockeys) including Grandmaster Flash, Afrika Bambaataa and DJ Kool Herc extended the breaks of funk records, a major new musical genre emerged that sought to isolate the percussion breaks of hit funk, disco and soul songs. As hip hop's popularity grew, performers began speaking ("rapping") in sync with the beats, and became known as MCs or emcees. The Herculoids, made up of Herc, Coke La Rock, and DJ Clark Kent, were the earliest to gain major fame. The Bronx is referred to in hip-hop slang as "The Boogie Down Bronx", or just "The Boogie Down". This was hip-hop pioneer KRS-One's inspiration for his thought provoking group BDP, or Boogie Down Productions, which included DJ Scott La Rock. Newer hip hop artists from the Bronx include Lord Toriq and Peter Gunz, Camp Lo, Swizz Beatz, Drag-On, Fat Joe, Terror Squad and Corey Gunz.[99]

Hush Hip Hop Tours has established a sightseeing tour of the Bronx showcasing the locations that helped shape hip hop culture and has the pioneers of hip hop as tour guides. The recent recognition of the Bronx as an important center of African-American culture, led Fordham University to establish the ongoing "Bronx African-American History Project (BAAHP)" [100].

The Bronx is the home of the New York Yankees, one of the leading baseball franchises. The original Yankee Stadium opened in 1923 on 161st Street and River Avenue, a year that saw the Yankeees bring home their first of 27 World Series Championships. With the famous facade, the short right field porch and Monument Park, Yankee Stadium has been home to many of baseball's greatest players including Babe Ruth, Lou Gehrig, Joe DiMaggio, Whitey Ford, Yogi Berra, Mickey Mantle, Reggie Jackson, Derek Jeter, Mariano Rivera and Alex Rodriguez.

The Stadium was the scene Lou Gehrig's Farewell Speech in 1939, Don Larsen's perfect game in the 1956 World Series, Roger Maris' record breaking 61st home run in 1961, and Reggie Jackson's 3 home runs to clinch Game 6 of the 1977 World Series. The original Yankee Stadium closed in 2008 to make way for a new Yankee Stadium in which the team started play in 2009; it's north-northeast of the 1923 Yankee Stadium, on the former site of Macombs Dam Park.

The Bronx is home to several Off-Off-Broadway theaters, many staging new works by immigrant playwrights from Latin America and Africa. The Pregones Theater [101], which produces Latin American work, opened a new 130-seat theater in 2005 on Walton Avenue in the South Bronx. Some artists from elsewhere in New York City have begun to converge on the area, and housing prices have nearly quadrupled in the area since 2002. However rising prices directly correlate to a housing shortage across the city and the entire metro area.

The Bronx Museum of the Arts, founded in 1971, exhibits 20th century and contemporary art through its central museum space and 11000 square feet (1000 m^2) of galleries. Many of its exhibitions are on themes of special interest to the Bronx. Its permanent collection features more than 800 works of art, primarily by artists from Africa, Asia and Latin America, including paintings, photographs, prints, drawings, and mixed media. The museum was temporarily closed in 2006 while it underwent a major expansion designed by the architectural firm Arquitectonica.

The Bronx has also become home to a peculiar poetic tribute, in the form of the Heinrich Heine Memorial, better known as the Lorelei Fountain from one of Heine's best-known works (1838) [102]. After Heine's German birthplace of Düsseldorf had rejected, allegedly for anti-Semitic motives, a centennial monument to the radical German-Jewish poet (1797–1856), his incensed German-American admirers, including Carl Schurz, started a movement to place one instead in Midtown Manhattan, at Fifth Avenue and 59th Street. However, this intention was thwarted by a combination of ethnic antagonism, aesthetic controversy and political struggles over the institutional control of public art.[103]

Lorelei Fountain in Joyce Kilmer Park overlooking the original Yankee Stadium.

In 1899, the memorial, by the Berlin sculptor Ernst Gustav Herter (1846–1917), finally came to rest, although subject to repeated vandalism, in the Bronx, at 164th Street and the Grand Concourse, or Joyce Kilmer Park near today's Yankee Stadium. (In 1999, it was moved to 161st Street and the Concourse.) In 2007, Christopher Gray of *The New York Times* described it as "a writhing composition in white Tyrolean marble depicting Lorelei, the mythical German figure, surrounded by mermaids, dolphins and seashells."[103]

One national landmark in the Bronx is the Hall of Fame for Great Americans, overlooking the Harlem River and designed by the renowned architect Stanford White. The never–landmarked Yankee Stadium, the "House that Ruth Built" and home to the New York Yankees since 1923, has been replaced with a similar-looking ballpark just across 161st Street.

The peninsular borough's maritime heritage is acknowledged in several ways.The City Island Historical Society and Nautical Museum [104] occupies a former public school designed by the New York City school system's turn-of-the-last-century master architect C. B. J. Snyder. The state's Maritime College in Fort Schuyler (on the southeastern shore) houses the Maritime Industry Museum.[105] In addition, the Harlem River is reemerging as "Scullers' Row" [106] due in large part to the efforts of the Bronx River Restoration Project,[107] a joint public-private endeavor of the city's parks department. Canoeing and kayaking on the borough's namesake river have been promoted by the Bronx River Alliance [108]. The river is also straddled by the New York Botanical Gardens, its neighbor, the Bronx Zoo, and a little further south, on the west shore, Bronx River Art Center [109].

The press and broadcasting

Newspapers The Bronx has several local newspapers, including *The Bronx News* [110], *Parkchester News*, *City News*, *The Riverdale Press*, *Riverdale Review*, *The Bronx Times Reporter*, *Inner City Press* [111] (which now has more of a focus on national issues) and *Co-Op City Times*. Four non-profit news outlets, *Norwood News*, *Mount Hope Monitor*, *Mott Haven Herald* and *The Hunts Point Express* serve the borough's poorer communities. The editor and co-publisher of *The Riverdale Press*, Bernard Stein, won the Pulitzer Prize for editorial writing for his editorials about Bronx and New York City issues in 1998. (Stein graduated from the Bronx High School of Science in 1959.)

The Bronx once had its own daily newspaper, *The Bronx Home News*, which started publishing on January 20, 1907 and merged into the *New York Post* in 1948. It became a special section of the Post, sold only in the Bronx, and eventually disappeared from view.

Radio and television One of New York City's major non-commercial radio broadcasters is WFUV, an National Public Radio–affiliated 50,000-watt station broadcasting from Fordham University's Rose Hill campus in the Bronx. The radio station's antenna is atop an apartment building owned by Montefiore Medical Center.

The City of New York has an official television station run by the NYC Media Group and broadcasting from Bronx Community College, and Cablevision operates News 12 The Bronx, both of which feature programming based in the Bronx. Co-op City was the first area in the Bronx, and the first in New York beyond Manhattan, to have its own cable television provider. The local Public-access television station BronxNet provides Government-access television (GATV) public affairs programming in addition to programming produced by Bronx residents.[112]

Screen, literature and song

On screen (film and television)

Middle 20th century movies set in the Bronx portrayed densely settled, working-class, urban culture. Hollywood films such as *From This Day Forward* (1946), set in Highbridge, occasionally delved into Bronx life. Paddy Chayefsky's Academy Award-winning *Marty* was the most notable examination of working class Bronx life[113] was also explored by Chayefsky in his 1956 film *The Catered Affair*, and in the 1993 Robert De Niro/Chazz Palminteri film, *A Bronx Tale*, Spike Lee's 1999 movie *Summer of Sam*, centered in an Italian-American Bronx community, 1994's *I Like It Like That* that takes place in the predominately Puerto Rican neighborhood of the South Bronx, and *Doughboys*, the story of two Italian-American brothers in danger of losing their bakery thanks to one brother's gambling debts.

The Bronx's gritty urban life had worked its way into the movies even earlier, with depictions of the "Bronx cheer", a loud flatulent-like sound of disapproval, allegedly first made by New York Yankees fans. The sound can be heard, for example, when Spike Jones sings "Der Fuehrer's Face" (from the 1942 Disney animated film of the same name), repeatedly lambasting Adolf Hitler with: "We'll Heil! (Bronx cheer) Heil! (Bronx cheer) Right in Der Fuehrer's Face!"[114]

Other movies have also used the term *Bronx* for comic effect, such as "Bronx", the character on the Disney animated series *Gargoyles*.

Starting in the 1970s, the Bronx often symbolized violence, decay, and urban ruin. The wave of arson in the South Bronx in the 1960s and 1970s inspired the observation that "The Bronx is burning": in 1974 it was the title of both a New York Times editorial and a BBC documentary film. The line entered the pop-consciousness with Game Two of the 1977 World Series, when a fire broke out near Yankee Stadium as the team was playing the Los Angeles Dodgers. Numerous fires had previously broken out in the Bronx prior to this fire. As the fire was captured on live television, announcer Howard Cosell stated, "There it is, ladies and gentlemen: the Bronx is burning". Historians of New York City frequently point to Cosell's remark as an acknowledgement of both the city and the borough's decline.[115] A new feature-length documentary film by Edwin Pagan called *Bronx Burning* [116] is in production[117] in 2006, chronicling what led up to the numerous arson-for-insurance fraud fires of the 1970s in the borough.

Bronx gang life was depicted in the 1974 novel *The Wanderers* by Bronx native Richard Price and the 1979 movie of the same name. They are set in the heart of the Bronx, showing apartment life and the then-landmark Krums ice cream parlor. In the 1979 film *The Warriors*, the eponymous gang go to a meeting in Van Cortlandt Park in the Bronx, and have to fight their way out of the borough and get back to Coney Island in Brooklyn. The 2005 video game adaptation features levels called Pelham, Tremont, and "Gunhill" (a play off the name Gun Hill Road).

This theme lends itself to the title of *The Bronx Is Burning*, an eight-part ESPN TV mini-series (2007) about the New York Yankees' drive to winning baseball's 1977 World Series. The TV series emphasizes the boisterous nature of the team, led by manager Billy Martin, catcher Thurman Munson and outfielder Reggie Jackson, as well as the malaise of the Bronx and New York City in general during that time, such as the blackout, the city's serious financial woes and near bankruptcy, the arson for insurance payments, and the election of Ed Koch as mayor.

The 1981 film *Fort Apache, The Bronx* is another film that used the Bronx's gritty image for its storyline. The movie's title is from the nickname for the 41st Police Precinct in the South Bronx which was nicknamed "Fort Apache". Also from 1981 is the horror film *Wolfen* making use of the rubble of the Bronx as a home for werewolf type creatures. *Knights of the South Bronx*, a true story of a teacher who worked with disadvantaged children, is another film also set in the Bronx released in 2005.

The Bronx was the setting for the 1983 film *Fuga dal Bronx*, also known as *Bronx Warriors 2* and *Escape 2000*, an Italian B-movie best known for its appearance on the television series Mystery Science Theatre 3000. The plot revolves around a sinister construction corporation's plans to depopulate, destroy and redevelop the Bronx, and a band of rebels who are out to expose the corporation's murderous ways and save their homes. The film is memorable for its almost incessant use of the phrase, "Leave the Bronx!" Many of the movie's scenes were filmed in Queens, substituting as the Bronx. *Rumble in the Bronx* was a 1995 Jackie Chan kung-fu film, another which popularised the Bronx to international audiences. **Last Bronx**, a 1996 Sega game played on the bad reputation of the Bronx to lend its name to an alternate version of post-Japanese bubble Tokyo, where crime and gang warfare is rampant.

Bronx born and reared Nancy Savoca's 1989 comedy *True Love*, explores two Italian-American Bronx sweethearts in the days before their wedding. The film, which debuted Annabella Sciorra and Ron Eldard as the betrothed couple, won the Grand Jury Prize at that year's Sundance Film Festival.

The CBS television sitcom *Becker*, 1998–2004, was more ambiguous. The show starred Ted Danson as Dr. John Becker, a doctor who operated a small practice and was constantly annoyed by his patients, co-workers, friends, and practically everything and everybody else in his world. It showed his everyday life as a doctor working in a small clinic in the Bronx.

Penny Marshall's 1990 film *Awakenings*, which was nominated for several Oscars, is based on neurologist Oliver Sacks' 1973 account of his psychiatric patients at Beth Abraham Hospital in the Bronx who were paralyzed by a form of encephalitis but briefly responded to the drug L-dopa. Robin Williams played the physician; Robert De Niro was one of the patients who emerged from a catatonic (frozen) state. The home of Williams' character was shot not far from Sacks' actual City Island residence. A 1973 Yorkshire Television documentary [118] and "A Kind of Alaska", a 1985 play by Harold Pinter,[119] were also based on Sacks' book.

Gus Van Sant's 2000 *Finding Forrester* was quickly billed "Good Will Hunting in the Hood." Sean Connery is in the title role of a reclusive old man who 50 years earlier wrote a single novel that garnered the Pulitzer Prize. He meets 16 year old Jamal – actor Rob Brown – a gifted, Bronx reared basketball player and aspiring writer, and becomes his mentor. The movie includes stock footage of Bronx housing projects from 1990, as well as some other scenes shot in Manhattan and Brooklyn.

The 2012 documentary "South Bronx United" features the Mott Haven neighborhood and its conflict over the online grocery delivery service Fresh Direct's move of their trucking facility from Long Island City to the South Bronx.

In literature

The Bronx has been featured significantly in fiction literature. All of the characters in Herman Wouk's City Boy: The Adventures of Herbie Bookbinder (1948) live in the Bronx, and about half of the action is set there. Kate Simon's *Bronx Primitive: Portraits of a Childhood* is directly autobiographical, a warm account of a Polish-Jewish girl in an immigrant family growing up before World War II, and living near Arthur Avenue and Tremont Avenue.[120] In Jacob M. Appel's short story, "The Grand Concourse" (2007),[121] a woman who grew up in the iconic Lewis Morris Building returns to the Morrisania neighborhood with her adult daughter. Similarly, in Avery Corman's book *The Old Neighborhood* (1980),[122] an upper-middle class white protagonist returns to his birth neighborhood (Fordham Road and the Grand Concourse), and learns that even though the folks are poor, Hispanic and African-American, they are good people.

By contrast, Tom Wolfe's *Bonfire of the Vanities* (1987)[123] portrays a wealthy, white protagonist, Sherman McCoy, getting lost off the Major Deegan Expressway in the South Bronx and having an altercation with locals. A substantial piece of the last part of the book is set in the resulting riotous trial at the Bronx County Court House. However, times change, and in 2007, the *New York Times* reported that "the Bronx neighborhoods near the site of Sherman's accident are now dotted with townhouses and apartments." In the same article, the Reverend Al Sharpton (whose fictional analogue in the novel is "Reverend Bacon") asserts that "twenty years later, the cynicism of *The Bonfire of the Vanities* is as out of style as Tom Wolfe's wardrobe."[124]

Don DeLillo's *Underworld* (1997) is also set in the Bronx and offers a perspective on the decline of the area from the 1950s onwards. John Patrick Shanley's "Savage in Limbo" is set in a 1980s Bronx bar called 'Scales' where the frustrated characters feel they are unable to move.

In poetry, the Bronx has been immortalized by one of the world's shortest couplets:

> The Bronx
>
> No Thonx
>
> Ogden Nash, *The New Yorker*, 1931

Nash repented 33 years after his calumny, penning in 1964 the following prose poem to the Dean of Bronx Community College:

> I can't seem to escape
>
> the sins of my smart-alec youth;
>
> Here are my amends.
>
> I wrote those lines, "The Bronx?
>
> No thonx";
>
> I shudder to confess them.
>
> Now I'm an older, wiser man
>
> I cry, "The Bronx?
>
> God bless them!"[38]

In song

Passing through: The theme song to the 1960s U.S. television comedy series *Car 54, Where Are You?* begins "There's a holdup in The Bronx."[125] And the song "New York, New York" (by Betty Comden and Adolph Green from the 1940s musical comedy and film, "On the Town") explains that "The Bronx is up and the Battery's down." Another song, "Manhattan" (by Richard Rodgers and Lorenz Hart for the 1925 musical "Garrick Gaieties"), declares "We'll have Manhattan,/The Bronx and Staten/Island too./It's lovely going through/the zoo."

Bronx Local: While hundreds of songs about New York City, Manhattan and Brooklyn can be found in Wikipedia's List of songs about New York City and also in Marc Ferris's 5-page, 15-column list of "Songs and Compositions Inspired by New York City" in *The Encyclopedia of New York City* (1995),[126] only a handful refer to The Bronx.

Ferris's extensive but selective 1995 list mentions only four songs referring specifically to The Bronx:

- "On the Banks of The Bronx" (1919), William LeBaron, Victor Jacobi
- "Bronx Express" (1922), Creamer and Layton
- "The Tremont Avenue Cruisewear Fashion Show" (1973), Jerry Livingston, Mark David
- "I Love the New York Yankees" (1987), Paula Lindstrom

But Wikipedia's own list also currently mentions:

- "Alfie From The Bronx" by The Toy Dolls
- "Back To The Bronx" by 2 Live Crew
- "Boogie Down Bronx" by JVC Force
- "Bronx" by Kurtis Blow
- "The Bronx" by Regina Spektor
- "Bronx Backyard" by The Johnny Seven Band
- "The Bronx Is Beautiful" by Robert Klein
- "Bronx Keeps Creating It" by Fat Joe
- "Bronx Tale" by Fat Joe
- "Bronx War Stories" by A.I.G.
- "Cousin in The Bronx" by Kaiser Chiefs
- "Cross Bronx Expressway" by Lord Tariq and Peter Gunz
- "Deja Vu (Uptown Baby)" by Lord Tariq & Peter Gunz
- "Ha Ya Doin? Yankees" – The Haya Doin'? Boys
- "Here Come the Yankees", by Bob Bundin and Lou Stallman
- "Jenny From The Block" by Jennifer Lopez featuring Styles P & Jadakiss
- "On The Streets Of The Bronx" by The Moonglows
- "Our Lady of The Bronx" by Black 47
- "Rockin' The Bronx" by Black 47
- "South Bronx" by Boogie Down Productions
- "Lost in the Flood" by Bruce Springsteen

See also

- Bronx gangs (1950s-1960s)
- Bronx cheer (gesture)
- Bronx Bombers
- Joseph P. Day, early land auctioneer
- List of people from the Bronx
- National Register of Historic Places listings in Bronx County, New York

General:

- List of counties in New York

References

[1] 2011 borough population estimates are taken from the annual database (http://www.census.gov/popest/data/counties/totals/2011/index.html) of county population estimates from the U.S. Census Bureau, specifically this Microsoft Excel spreadsheet for New York state's counties, Table 1. Annual Estimates of the Resident Population for Counties of New York: April 1, 2010 to July 1, 2011 (http://www.census.gov/popest/data/counties/totals/2011/tables/CO-EST2011-01-36.xls), retrieved on May 13, 2012

[2] U.S. Census Bureau, *County and City Data Book:2007* (http://www.census.gov/statab/ccdb/ccdbstcounty.html) Table B-1, Area and Population, retrieved on July 12, 2008. New York County (Manhattan) was the nation's densest-populated county, followed by Kings County (Brooklyn), Bronx County, Queens County and San Francisco, California.

[3] http://bronxboropres.nyc.gov/

[4] While the Bronx has an area of only 42 square miles (109 km^2), it has more residents than the 665000 square miles (1722342.093 km^2) of Alaska and Wyoming combined, according to Table 348 of the *Statistical Abstract of the United States: 2008* (http://www.census.gov/compendia/statab/cats/geography_environment.html) Ranked areas of the boroughs from U.S. Census Bureau, *County and City Data Book:2007* (http://www.census.gov/statab/ccdb/ccdbstcounty.html) Table B-1, Area and Population, retrieved on July 13, 2008.

[5] Lloyd Ultan, Bronx Borough Historian, "History of the Bronx River," (http://www.bronxriver.org/?pg=content&p=abouttheriver&m1=9&m2=58) Paper presented to the Bronx River Alliance, November 5, 2002 (notes taken by Maarten de Kadt, November 16, 2002), retrieved on August 29, 2008. This 2½ hour talk covers much of the early history of the Bronx as a whole, in addition to the Bronx River.

[6] On the start of business for Bronx County: Bronx County In Motion. New Officials All Find Work to Do on Their First Day. (http://query.nytimes.com/mem/archive-free/pdf?_r=2&res=9C05E7DA1F3BE633A25750C0A9679C946596D6CF) *The New York Times*, January 3, 1914 (PDF retrieved on June 26, 2008):

> "Despite the fact that the new Bronx County Court House is not completed there was no delay yesterday in getting the court machinery in motion. All the new county officials were on hand and the County Clerk, the District Attorney, the Surrogate, and the County Judge soon had things in working order. The seal to be used by the new county was selected by County Judge Louis D. Gibbs. It is circular. In the centre is a seated figure of Justice. To her right is an American shield and over the figure is written 'Populi Suprema.' ..."
>
> "Surrogate George M. S. Schulz, with his office force, was busy at the stroke of 9 o'clock. Two wills were filed in the early morning, but owing to the absence of a safe they were recorded and then returned to the attorneys for safe keeping. ..."
>
> "There was a rush of business to the new County Clerk's office. Between seventy-five and a hundred men applied for first naturalization papers. Two certificates of incorporation were issued, and seventeen judgments, seven lis pendens, three mechanics' liens and one suit for negligence were filed."
>
> "Sheriff O'Brien announced several additional appointments."

[7] Ladies and gentlemen, the Bronx is blooming! (http://news.yahoo.com/s/ap_travel/20080630/ap_tr_ge/travel_trip_wild_green_bronx) by Beth J. Harpaz, Travel Editor of The Associated Press (AP), June 30, 2008, retrieved on July 11, 2008

[8] "Jonas Bronx" (http://www.bronxhistoricalsociety.org/notebx.html). *Bronx Notables*. Bronx Historical Society. . Retrieved 2012-01-20.

[9] Wylie, Jonathon (1987), *The Faroe Islands: Interpretations of History* (http://books.google.nl/books?id=7kmEYtkttx4C&printsec=frontcover&dq=isbn:9780813115788&redir_esc=y#v=onepage&q&f=false), University of Kentucky Press, p. 209, ISBN 978-0-8131-1578-8, , "Jónas Bronck (or Brunck) was the son of Morten Jespersen Bronck.....Jónas seems to have gone to school in Roskilde in 1619, but found his way to Holland where he joined an expedition to Amsterdam."

[10] van Laer, A. J. F. (October 1916), *The American Historical Review*, **22, n.1**, Chicago: The University of Chicago Press on behalf of the American Historical Association, pp. 164–166, JSTOR 1836219, "… Jonas Bronck was a Dane …"

[11] Burrows, Edwin G.; Wallace, Mike (Michael L.) (1999), *Gotham, A History of New York City to 1898*, **1**, Oxford, New York: Oxford University Press, pp. 30–37, ISBN 0-19-511634-8 "…many of these colonists, perhaps as many as half of them, represented the same broad mixture of nationalities as New Amsterdam itself. Among them were Swedes, Germans, French, Belgians, Africans, and Danes (such as a certain Jonas Bronck)..."

[12] Van Rensselaer, Mariana Griswold (1909), *History of the city of New York in the seventeenth century* (http://www.worldcat.org/title/history-of-the-city-of-new-york-in-the-seventeenth-century/oclc/649654938?title=&detail=&page=frame&url=http://books.google.com/books?id=i9ET9y_FpgQC&checksum=9a3e1cdde2061df1e906bed9b5c04bae&linktype=digitalObject), **1**, New York: The Macmillan Company, p. 161,

[13] Braver (1998)

[14] *The Almanac of American Politics 2008*, edited by Michael Barone with Richard E. Cohen and Grant Ujifusa, National Journal Group, Washington, D.C., 2008 ISBN 978-0-89234-117-7 (paperback) or –116-0 (hardback), chapter on New York state

[15] U.S. Census Bureau, *Statistical Abstract of the United States: 2003*, Section 31, Table 1384. Congressional District Profiles — 108th Congress: 2000

[16] See the "Historical Populations" table in History above and its sources.

[17] American Fact Finder (U.S. Census Bureau): New York by County - Table GCT-PH1. Population, Housing Units, Area, and Density: 2000 Data Set: Census 2000 Summary File 1 (SF 1) 100-Percent Data (http://factfinder.census.gov/servlet/GCTTable?_bm=y&-context=gct&-ds_name=DEC_2000_SF1_U&-CONTEXT=gct&-mt_name=DEC_2000_SF1_U_GCTPH1_US9&-tree_id=4001&-redoLog=true&-_caller=geoselect&-geo_id=04000US36&-format=ST-2|ST-2S&-_lang=en), retrieved on February 6, 2009

[18] Olmsted (1989); Olmsted (1998)

[19] "Piano Workers May Strike" (http://query.nytimes.com/mem/archive-free/pdf?res=F20712FE3B5910738DDDA00A94D0405B898DF1D3). *The New York Times*. August 29, 1919. . Retrieved 25 January 2011.

[20] Christopher Gray, "Streetscapes: The New York Coliseum; From Auditorium To Bus Garage to..." (http://query.nytimes.com/gst/fullpage.html?res=9E0CE5DE103DF931A15750C0A964958260) *The New York Times*, Real Estate section, March 22, 1992, retrieved on July 2, 2008

[21] *The World Almanac and Book of Facts, 1943*, page 494, citing the American Jewish Committee and the Jewish Statistical Bureau of the Synagogue Council of America

[22] Remembrance of Synagogues Past: The Lost Civilization of the Jewish South Bronx (http://www.bronxsynagogues.org/ic/bronxsyn/survey.html), by Seymour J. Perlin, Ed.D. (retrieved on August 10, 2008), citing population estimates in "The Jewish Community Study of New York: 2002", UJA [United Jewish Appeal] Federation of New York, June 2004, and his own survey of synagogue sites.

[23] Alfred A. Knopf, New York, 1974; ISBN 0-394-72024-5

[24] Roderick Wallace: "A synergism of plagues: 'planned shrinkage,' contagious housing destruction, and AIDS in the Bronx." *Environmental Research*, October 1988, Vol. 47, No. 1, pp. 1–33, and "Urban desertification, public health and public order: 'planned shrinkage', violent death, substance abuse and AIDS in the Bronx", *Social Science & Medicine*, Vol. 37, No. 7 (1990) pp. 801–813 — abstracts retrieved on July 5, 2008 from PubMed (http://www.ncbi.nlm.nih.gov). One sentence in the abstract of the 1990 article reads, "Empirical and theoretical analyses strongly imply present sharply rising levels of violent death, intensification of deviant behaviors implicated in the spread of AIDS, and the pattern of the AIDS outbreak itself, have been gravely affected, and even strongly determined, by the outcomes of a program of 'planned shrinkage' directed against African-American and Hispanic communities, and implemented through systematic and continuing denial of municipal services—particularly fire extinguishment resources—essential for maintaining urban levels of population density and ensuring community stability."

[25] Issues such as redlining, hospital quality and what looked like the planned shrinkage of garbage collection became the contentious issues that sparked the Puerto Rican activists known as the Young Lords. The Young Lords coalesced with similar groups fighting for neighborhood empowerment, such as the Black Panthers, to protest urban renewal and arson for profit with sit-ins and marches. See pages 6–9 of the guide to *¡Palante Siempre Palante!* The Young Lords (http://www.pbs.org/pov/utils/youthviews/pov_palante_toolkit.pdf) a "P.O.V." (Point of View) documentary on the Public Broadcasting Service.

[26] For an example of this argument, as well as of several other theses mentioned here, see "When the Bronx was burning" (http://www.city-data.com/forum/new-york-city/257896-when-bronx-burning-6.html) *City-data* forum (blog), 2007, where rubygreta writes:"Rent control destroyed the Bronx, especially starting in the 1960s and 1970s, when oil prices rose through the roof, and heavily subsidized Coop City opened in the East Bronx. Essentially, tenants never moved out of their apartments because they had below-market rents thanks to rent control. The apartments deteriorated and common areas deteriorated because the landlords had no cash-flow. And no cash flow meant that they could not get mortgages for major repairs such as boilers, roofs and window replacement."

[27] "Arson for Hate and Profit" (http://www.time.com/time/magazine/article/0,9171,945795-2,00.html). *Time*. 1977-10-31. . Retrieved 2008-03-14.

[28] Gonzalez (2004)

[29] PERSPECTIVES: The 10-Year Housing Plan; Issues for the 90's: Management and Costs (http://query.nytimes.com/gst/fullpage.html?res=9C0CE5D6103AF934A35752C0A966958260), *The New York Times*, January 7, 1990

[30] *Neighborhood Change and the City of New York's Ten-Year Housing Plan* (http://www.fanniemaefoundation.org/programs/hpd/pdf/hpd_1004_ryzin.pdf) *Housing Policy Debate* • Volume 10, Issue 4. Fannie Mae Foundation 1999.

[31] NOS QUEDAMOS/WE STAY Melrose Commons, Bronx, New York (http://www.sustainable.org/casestudies/SIA_PDFs/SIA_New_York.pdf) Sustainable Communities Network Case Studies *Sustainability in Action* 1997, retrieved on July 6, 2008

[32] David Gonzalez, Yolanda Garcia, 53, Dies; A Bronx Community Force (http://www.nytimes.com/2005/02/19/obituaries/19garcia.html), *The New York Times*, February 19, 2005, retrieved on July 6, 2008

[33] Meera Subramanian, HOMES AND GARDENS IN THE SOUTH BRONX (http://journalism.nyu.edu/portfolio/subramanian/SBxGardens.html), *Portfolio*, November 8, 2005, New York University Department of Journalism, retrieved on July 6, 2008

[34] Wealthy are drowning in new bank branches, says study (http://www.nydailynews.com/news/2007/09/10/2007-09-10_wealthy_are_drowning_in_new_bank_branche.html), *New York Daily News*, Monday, September 10, 2007

[35] Superintendent Neiman Addresses the Ninth Annual Bronx Bankers Breakfast (http://www.banking.state.ny.us/sp070615.htm) June 15, 2007. Among the remarks of Richard H. Neiman, New York State's Superintendent of Banks, were these: "The Bronx was an economically stable community until the mid 1960s when the entire South Bronx struggled with major construction, real estate issues, red-lining, and block busting. This included a thoroughfare that divided communities, the deterioration of property as a result of rent control, and decrease in the value of real estate.

Thanks to strong community leadership, advances in policing, social services, and changing economic migration patterns to New York City, the Bronx is undergoing a resurgence, with new housing developments and thriving business. From 2000 to 2006, there was a 2.2% increase in population and home ownership rates increased by 19.6%. "When I look at maps of the Bronx, it's not difficult to see the areas that don't have bank branches. These areas, which are prime locations for new bank branches, include Community districts 1, 3, 4, 5, 6, 9, and 12."

[36] New bank targets Latinos in South Bronx (http://www.hispanicmpr.com/2007/12/11/new-bank-targets-latinos-in-south-bronx/) December 11, 2007

[37] On June 30, 2005, there were 129 Federally insured banking offices in the Bronx, for a ratio of 1.0 offices for every 10,000 inhabitants. By contrast the national financial center of Manhattan had 555 for a ratio of 3.5/10,000, Staten Island a ratio of 1.9, Queens 1.7 and Brooklyn 1.1. In New York State as a whole the ratio was 2.6 and in the United States, 3.5 (a single office can serve more people in a more-densely-populated area.) U.S. Census Bureau, *City and County Data Book, 2007* Table B-11. Counties – Banking, Retail Trade, and Accommodation and Food Services (http://www.census.gov/statab/ccdb/cc07_tabB11.xls) For 1997 and 2007, Federal Deposit Insurance Corporation, Summary of Deposits; summary tables (http://www4.fdic.gov/SOD/sodSummary.asp?barItem=3) Deposits of all FDIC-Insured Institutions Operating in New York: State Totals by County — all retrieved on July 15–16, 2008.

[38] Williams, Timothy (2006-06-27). "Celebrities Now Give Thonx for Their Roots in the Bronx" (http://www.nytimes.com/2006/06/27/nyregion/27bronx.html). *The New York Times*. . Retrieved 2008-03-14.

[39] Topousis, Tom (2007-07-23). "Bx is Booming" (http://www.nypost.com/seven/07232007/news/regionalnews/bx__is_booming_regionalnews_tom_topousis.htm). *New York Post*. . Retrieved 2008-03-15.

[40] FUTURE OF NEW WARDS; New-York's Possession in Westchester County Rapidly Developing. (http://query.nytimes.com/gst/abstract.html?res=9F02EED6123EE333A25754C1A9639C94679ED7CF) *The New York Times*, Wednesday, May 17, 1896, page 15 (The subheadlines continue "TROLLEY AND STEAM ROAD SYSTEMS Vast Areas Being Brought Close to the Heart of the City – Miles of New Streets and Sewers. BOTANICAL AND ZOOLOGICAL GARDENS. Advantages That Will Soon Relieve Crowded Sections of the City of Thousands of Their Inhabitants.") This is a very useful glimpse into the state of the Bronx (and the hopes of Manhattan's pro-Consolidation forces) as parks, housing and transit were all being rapidly developed.

[41] "Find a County" (http://www.naco.org/Counties/Pages/FindACounty.aspx). National Association of Counties. . Retrieved 2011-06-07.

[42] "NYPL Map Warper: Browse Maps" (http://maps.nypl.org/warper/mapscans?field=title&query=bronx&show_warped=0). Maps.nypl.org. . Retrieved 2012-11-07.

[43] Berger, Joseph (2010-07-19). "Reclaimed Jewel Whose Attraction Can Be Perilous" (http://www.nytimes.com/2010/07/20/nyregion/20river.html?ref=nyregion). *New York Times*. . Retrieved 2010-07-21.

[44] Bronx High Point (http://www.peakbagger.com/peak.aspx?pid=7110) and Ascent of Bronx Point on 2008-6-24 (http://www.peakbagger.com/climber/ascent.aspx?aid=33535) at Peakbaggers.com, retrieved on July 22, 2008

[45] U.S. Census 2000 Gazetteer Files (http://www.census.gov/geo/www/gazetteer/places2k.html#cousub) retrieved n July 26, 2008

[46] Waterfront Development Initiative (http://bronxboropres.nyc.gov/en/gv/president/waterfront.htm), Bronx Borough President's office, March 19, 2004, retrieved on July 29, 2008

[47] Last Section Of Macombs Dam Park Closes To The Public For Redevelopment *On-site construction begins on Garage A and the New Macombs Dam Park* (http://www.nycgovparks.org/sub_newsroom/press_releases/press_releases.php?id=19967), Press Release, November 1, 2007, New York City Department of Parks and Recreation retrieved on July 19, 2008

[48] http://www.nycgovparks.org/sub_your_park/your_park.html

[49] In September 2008, Fordham University and its neighbor, the Wildlife Conservation Society, a global research organization which operates the Bronx Zoo, will begin a joint program leading to a Master of Science degree in adolescent science education (biology grades 7–12).

[50] Crotona Park (http://nycgovparks.org/parks/crotonapark) New York City Department of Parks and Recreation, retrieved on July 20, 2008

[51] Article on the Bronx by Gary D. Hermalyn and Lloyd Ultan in *The Encyclopedia of New York City* (1995 – see Further reading for bibliographic details)

[52] Jerome Park (http://www.nycgovparks.org/sub_your_park/historical_signs/hs_historical_sign.php?id=11042) (New York City Department of Parks and Recreation, retrieved on July 12, 2008).

[53] Bronx Parks for the 21st Century (http://nycgovparks.org/sub_your_park/croton/html/main_page.html), New York City Department of Parks and Recreation, retrieved on July 20, 2008. This links to both an interactive map and a downloadable (1.7 MB PDF) map showing nearly every public park and green space in the Bronx.

[54] As Maps and Memories Fade, So Do Some Bronx Boundary Lines (http://www.nytimes.com/2006/09/16/nyregion/16bronx.html) by Manny Fernandez, *The New York Times*, September 16, 2006, retrieved on August 3, 2008

[55] Most correlations with Community Board jurisdictions in this section come from Bronx Community Boards (http://www.bronxmall.com/commboards/index.html) at the Bronx Mall web-site, and New York: a City of Neighborhoods (http://www.nyc.gov/html/dcp/html/neighbor/neigh.shtml), New York City Department of City Planning, both retrieved on August 5, 2008

[56] FIELDSTON PROPERTY OWNERS' ASSOCIATION, INC. BY-LAWS (http://fpoa.info/images/FPOA_APPROVED_BY-LAWS.doc), by the FPOA, September 17, 2006

[57] The Hub (http://www.forgotten-ny.com/STREET SCENES/hub/hub.html)

[58] Bronx Neighborhood Histories (http://www.lehman.edu/vpadvance/artgallery/publicart/neighborhood.htm)

[59] Bronx Hub revival gathers steam (http://www.therealdeal.net/issues/JULY_2006/1151609191.php)

[60] Bronx Hub (http://www.sorkinstudio.com/Bronx Hub.htm)

[61] "Bronx Terminal Market - Wikipedia, the free encyclopedia" (http://en.wikipedia.org/wiki/Gateway_Center_at_Bronx_Terminal_Market). En.wikipedia.org. . Retrieved 2012-11-07.

[62] "Chains of Silver: Gateway Center At Bronx Terminal Market Earns LEED Silver Bona Fides"

[63] Bronx factsheet (http://www.tstc.org/reports/cpsheets/Bronx_factsheet.pdf)

[64] " NYC Post Offices to observe Presidents' Day (http://www.usps.com/communications/newsroom/localnews/ny/2009/ny_2009_0211.htm)." *United States Postal Service*. February 11, 2009. Retrieved on May 5, 2009.

[65] " Post Office Location – BRONX GPO (http://usps.whitepages.com/service/post_office/33768?p=1&s=ny&service_name=post_office&z=10451)." *"United States Postal Service*. Retrieved on May 5, 2009.

[66] "Photos: Bronx Residents on Obama" (http://www.newsweek.com/photo/2009/01/17/photos-bronx-residents-on-obama.html). *Newsweek*. Jan. 17, 2009. . Retrieved 12 May 2011.

[67] "New York - Race and Hispanic Origin for Selected Cities and Other Places: Earliest Census to 1990" (http://www.census.gov/population/www/documentation/twps0076/twps0076.html). U.S. Census Bureau. . Retrieved May 4, 2012.

[68] Quick Table QT-P9. Hispanic or Latino by Type: 2000; Data Set: Census 2000 Summary File 1 (SF 1) 100-Percent Data; Geographic Area: Bronx County, New York (http://factfinder.census.gov) drawn from U.S. Census Bureau, Census 2000 Summary File 1, Matrix PCT11, retrieved on August 7, 2008

[69] Census Estimates for New York April 1, 2010 to July 1, 2011 http://www.census.gov/popest/data/cities/totals/2011/files/SUB-EST2011_36.csv

[70] (1) *Population 1790–1960: The World Almanac and Book of Facts* 1966, page 452, citing estimates of the Department of Health, City of New York.

(2) *Population 1790–1990:* Article on "population" by Nathan Kantrowitz in *The Encyclopedia of New York City*, edited by Kenneth T. Jackson (Yale University Press, 1995 ISBN 0-300-05536-6), citing the United States Census Bureau

N.B., Estimates in (1) and (2) before 1920 re-allocate the Census population from the counties whose land is now partly occupied by Bronx County.

(3) *Population 1920–1990:* Population of Counties by Decennial Census: 1900 to 1990, Compiled and edited by Richard L. Forstall, Population Division, US Bureau of the Census (http://www.census.gov/population/cencounts/ny190090.txt), United States Census Bureau, Washington, D.C. 20233, March 27, 1995, retrieved July 4, 2008.

[71] "American FactFinder" (http://factfinder.census.gov/servlet/ADPTable?_bm=y&-geo_id=05000US36005&-qr_name=ACS_2009_1YR_G00_DP5&-context=adp&-ds_name=&-tree_id=309&-_lang=en&-redoLog=false&-format=). Factfinder.census.gov. . Retrieved 2012-11-07.

[72] "American FactFinder" (http://factfinder.census.gov/servlet/ADPTable?_bm=y&-context=adp&-ds_name=ACS_2009_1YR_G00_&-tree_id=309&-redoLog=true&-_caller=geoselect&-geo_id=05000US36005&-format=&-_lang=en). Factfinder.census.gov. . Retrieved 2012-11-07.

[73] "American FactFinder" (http://factfinder.census.gov/servlet/ADPTable?_bm=y&-geo_id=05000US36005&-qr_name=ACS_2007_3YR_G00_DP3YR5&-context=adp&-ds_name=&-tree_id=3307&-_lang=en&-redoLog=false&-format=). Factfinder.census.gov. . Retrieved 2012-11-07.

[74] "American FactFinder" (http://factfinder.census.gov/servlet/ADPTable?_bm=y&-geo_id=05000US36005&-qr_name=ACS_2007_3YR_G00_DP3YR2&-context=adp&-ds_name=&-tree_id=3307&-_lang=en&-redoLog=false&-format=). Factfinder.census.gov. . Retrieved 2012-11-07.

[75] Historical Census Browser (http://fisher.lib.virginia.edu/collections/stats/histcensus/index.html) University of Virginia, Geospatial and Statistical Data Center, retrieved on August 7, 2008, querying 1930 Census for New York State. "The data and terminology presented in the Historical Census Browser are drawn directly from historical volumes of the U.S. Census of Population and Housing."

[76] Oscar Johnson, "Chilly Coexistence: Africans and African Americans in the Bronx" (http://www.columbia.edu/itc/journalism/gissler/anthology/Chill-Johnson.html), *Race Anthology* (Columbia University journalism program), Spring 2000, retrieved on August 7, 2008. According to this story, U.S. Immigration and Naturalization Service data show that in 1996, about two-thirds of those Ghanaians visiting the United States, and nearly three-fourths of those naturalized, arrived in New York City.

[77] Quick Tables QT-P15 and QT-P22 (http://factfinder.census.gov/servlet/QTTable?_bm=y&-state=qt&-context=qt&-qr_name=DEC_2000_SF3_U_QTP15&-qr_name=DEC_2000_SF3_U_QTP22&-ds_name=DEC_2000_SF3_U&-CONTEXT=qt&-tree_id=402&-all_geo_types=N&-redoLog=true&-geo_id=05000US36005&-search_results=01000US&-format=&-_lang=en), U.S. Census Bureau, retrieved on August 10, 2008

[78] "American FactFinder" (http://factfinder.census.gov). United States Census Bureau. . Retrieved 2008-01-31.

[79] Kappstatter, Bob, (2/18/09), *Bronx Beep Bound for D.C.*, New York Daily News

[80] Trymaine Lee, *Borough Voters Elect Díaz as New Borough President* (http://www.nytimes.com/2009/04/22/nyregion/22bronx.html), The New York Times, *New York edition, April 22, 2009, page A24, retrieved on May 13, 2009*

[81] Kappstatter, Bob, (4/22/09), *Ruben Diaz Cruises to Victory in Bronx Borough President Special Election*, New York Daily News

[82] Board of Elections in the City of New York, Bronx Borough President special election results, April 21, 2009 (http://www.vote.nyc.ny.us/pdf/results/2009/BronxBoroPresident_4_21_09/BronxBoroPresident-Recap.pdf) (PDF with details by Assembly District, April 29, 2009), retrieved on May 13, 2009

[83] The Role of African American Males in Politics and Government. T.D. Snipe. *Annals*, 2000 - litigation-essentials.lexisnexis.com

[84] New York State Board of Elections: 2006 Results Page (http://www.elections.ny.gov/2006ElectionResults.html), retrieved on July 23, 2008.

[85] Board of Elections in the City of New York (http://www.vote.nyc.ny.us/results.html) Summary of Election Results (1999–2008), retrieved on July 21, 2008.

[86] *The World Almanac and Book of Facts* for 1929 & 1957; Our Campaigns (New York Counties Bronx President History) (http://ourcampaigns.com/ContainerHistory.html?ContainerID=9099);*The Encyclopedia of New York City* (see Further reading below), article on "government and politics"

[87] (The Republican line exceeded the ALP's in every other borough)

[88] To see a comparison of borough votes for Mayor, see New York City mayoral elections#How the boroughs voted

[89] QT-P19. School Enrollment: 2000; Data Set: Census 2000 Summary File 3 (SF 3) – Sample Data; Geographic Area: Bronx County, New York (http://factfinder.census.gov/servlet/QTTable?_bm=y&-context=qt&-qr_name=DEC_2000_SF3_U_QTP19&-ds_name=DEC_2000_SF3_U&-CONTEXT=qt&-tree_id=403&-all_geo_types=N&-geo_id=05000US36005&-search_results=01000US&-format=&-_lang=en), U.S. Census Bureau, retrieved August 22, 2008

[90] U.S. Census Bureau, *County and City Data Book:2007* (http://www.census.gov/statab/ccdb/ccdbstcounty.html), Table B-4. Counties – Population Characteristics

[91] Chronopoulos, Themis. ""Urban Decline and the Withdrawal of New York University from University Heights, The Bronx." The Bronx County Historical Society Journal XLVI (Spring/Fall 2009): 4-24." (http://themis.slass.org/university-heights.html). . Retrieved 6/12/2012.

[92] Monroe College history (http://www.monroecollege.edu/aboutmonroe/monroeshistory) (from the College's web site) retrieved on July 27, 2008.

[93] 2007 Fort Greene Park Summer Literary Festival website (http://www.nywriterscoalition.org/litfest.htm). See also the Flickr.com photograph album of the 2007 Festival (http://www.flickr.com/photos/nywc/sets/72157601898513545/)

[94] Edgar Allan Poe Cottage (http://www.bronxhistoricalsociety.org/about/poecottage.html), accessed October 9, 2006 Archived (http://web.archive.org/20061005212044/http://www.bronxhistoricalsociety.org/about/poecottage.html) October 5, 2006 at the Wayback Machine

[95] http://www.imdb.com/title/tt1169817/

[96] David Gonzalez, "Will Gentrification Spoil the Birthplace of Hip-Hop?" (http://www.nytimes.com/2007/05/21/nyregion/21citywide.html), *The New York Times*, May 21, 2007, retrieved on July 1, 2008

[97] Jennifer Lee, "Tenants Might Buy the Birthplace of Hip-Hop" (http://cityroom.blogs.nytimes.com/2008/01/15/tenants-might-buy-the-birthplace-of-hip-hop), *The New York Times*, January 15, 2008, retrieved on July 1, 2008

[98] Tukufu Zuberi ("detective"), *BIRTHPLACE OF HIP HOP*, History Detectives, Season 6, Episode 11, New York City, found at PBS official website (http://www.pbs.org/opb/historydetectives/investigations/611_hiphop.html). Accessed February 24, 2009.

[99] Johan Kugelberg, *Born in the Bronx* (http://www.rizzoliusa.com/catalog/display.pperl?isbn=9780789315403); New York: Rizzoli (Universe), 2007; ISBN 978-0-7893-1540-3.

[100] http://www.fordham.edu/academics/programs_at_fordham_/bronx_african_americ/index.asp

[101] http://www.pregones.org/

[102] http://www.business.uiuc.edu/vock/poetry/lorelei.html

[103] Christopher Gray, "Sturm und Drang Over a Memorial to Heinrich Heine" (http://www.nytimes.com/2007/05/27/realestate/27scap.html), *The New York Times*, May 27, 2007, retrieved on July 3, 2008. Archived on 2012-07-12 (http://www.webcitation.org/696vIk18P)

See also Public Art in the Bronx: Joyce Kilmer Park (http://www.lehman.edu/vpadvance/artgallery/publicart/J_kilmer_pk.htm), from Lehman College

[104] http://www.cityislandmuseum.org/

[105] Maritime Industry Museum (http://www.sunymaritime.edu/Maritime Museum/index.aspx), retrieved on August 21, 2008

[106] http://www.harlemrivercommunityrowing.org/

[107] http://www.bronxriver.org/puma/images/usersubmitted/greenway_plan/

[108] http://www.bronxriver.org/

[109] http://www.bronxriverart.org

[110] http://www.bxnews.net

[111] "(some) About Us" (http://www.innercitypress.org/aboutus.html). Inner City Press. . Retrieved 2012-11-07.

[112] Its website showcases very short selections (less than 20 seconds and over 2 MB each in uncompressed AIFF format) from *Bronx Music Vol.1* (http://www.bronxnet.org/info/music/bxmusic.htm), an out-of-press compact disc of the old and new sounds and artists of the Bronx.

[113] Chronopoulos, Themis. ""Paddy Chayefsky's 'Marty' and Its Significance to the Social History of Arthur Avenue, The Bronx, in the 1950s." The Bronx County Historical Society Journal XLIV (Spring/Fall 2007): 50-59." (http://themis.slass.org/marty.html). .

[114] David Hinkley, "Scorn and disdain: Spike Jones giffs Hitler der old birdaphone, 1942." *New York Daily News*,"March 3, 2004.http://www.nydailynews.com/archives/news/2004/03/03/2004-03-03_scorn_and_disdain_spike_jone.html

[115] Mahler, Jonathan (2005). *Ladies and Gentlemen, the Bronx is Burning*. Farrar, Straus and Giroux. ISBN 0-312-42430-2.

[116] http://www.imdb.com/title/tt0906735/

[117] "Opportunities for Arts Organizations and Community Based Organizations" (http://www.bronxarts.org/newsletter/200601.html). *E-News Update*. Bronx Council on the Arts. January 2006. .
[118] http://query.nytimes.com/gst/fullpage.html?res=9A00E3D8103DF932A15752C0A9679C8B63
[119] (ISBN 0-573-12129-X)
[120] Kate Simon, *Bronx Primitive: Portraits in a Childhood.* New York: Harper Colophon, 1983.
[121] *The Threepenny Review*, Volume 109, Spring 2007 (http://www.threepennyreview.com/tocs/109_sp07.html)
[122] Avery Corman, *The Old Neighborhood*, Simon and Schuster, 1980; ISBN 0-671-41475-5
[123] Tom Wolfe, *The Bonfire of the Vanities*, Farrar, Straus and Giroux 1987 (hardback) ISBN 978-0-374-11535-7, Picador Books 2008 (paperback) ISBN 978-0-312-42757-3
[124] Anne Barnard, Twenty Years After 'Bonfire,' A City No Longer in Flames (http://www.nytimes.com/2007/12/10/nyregion/10bonfire.html), *The New York Times*, December 10, 2007, retrieved on July 1, 2008
[125] Car 54, Where Are You?#Theme song
[126] *The Encyclopedia of New York City*, edited by Kenneth T. Jackson (Yale University Press and The New-York Historical Society, New Haven, Connecticut, 1995 ISBN 0-300-05536-6), pages 1091–1095

Further reading

- Barrows, Edward, and Mike Wallace. *Gotham: A History of New York City to 1898* (1999)
- Baver, Sherrie L. "Development of New York's Puerto Rican Community", *Bronx County Historical Society Journal* 1988 25(1): 1–9
- Briggs, Xavier de Souza, Anita Miller and John Shapiro. 1996. "CCRP in the South Bronx." Planners' Casebook, Winter.
- Chronopoulos, Themis. “Paddy Chayefsky's ‘Marty’ and Its Significance to the Social History of Arthur Avenue, The Bronx, in the 1950s." The Bronx County Historical Society Journal XLIV (Spring/Fall 2007): 50-59.
- Chronopoulos, Themis. “Urban Decline and the Withdrawal of New York University from University Heights, The Bronx." The Bronx County Historical Society Journal XLVI (Spring/Fall 2009): 4-24.
- de Kadt, Maarten. *The Bronx River: An Environmental and Social History.* The History Press (2011)
- DiBrino, Nicholas. *The History of the Morris Park Racecourse and the Morris Family* (1977)
- Federal Writers' Project. *New York City Guide: A Comprehensive Guide to the Five Boroughs of the Metropolis: Manhattan, Brooklyn, the Bronx, Queens, and Richmond* (1939) online edition (http://www.questia.com/read/99253940?title=New York City Guide: A Comprehensive Guide to the Five Boroughs of the Metropolis: Manhattan, Brooklyn, the Bronx, Queens, and Richmond)
- Gonzalez, Evelyn. *The Bronx.* (Columbia University Press, 2004. 263 ISBN 0-231-12114-8), scholarly history focused on the slums of the South Bronx online edition (http://www.questia.com/read/114330210?title=The Bronx)
- Goodman, Sam. "The Golden Ghetto: The Grand Concourse in the Twentieth Century", *Bronx County Historical Society Journal* 2004 41(1): 4–18 and 2005 42(2): 80–99
- Jackson, Kenneth T., ed. *The Encyclopedia of New York City*, (Yale University Press and The New-York Historical Society, (1995) ISBN 0-300-05536-6), has entries, maps, illustrations, statistics and bibliographic references on almost all of the significant topics in this article, from the entire borough to individual neighborhoods, people, events and artistic works.
- Jonnes, Jull. *South Bronx Rising: The Rise, Fall, and Resurrection of an American City* (2002) online edition (http://www.questia.com/read/111733280?title=South Bronx Rising: The Rise, Fall, and Resurrection of an American City)
- McNamara, John *History In Asphalt: The Origin of Bronx Street and Place Names* (1993) ISBN 0-941980-16-2
- McNamara, John *McNamara's Old Bronx* (1989) ISBN 0-941980-25-1
- Olmsted, Robert A. "A History of Transportation in the Bronx", *Bronx County Historical Society Journal* 1989 26(2): 68–91
- Olmsted, Robert A. "Transportation Made the Bronx", *Bronx County Historical Society Journal* 1998 35(2): 166–180

- Rodríguez, Clara E. *Puerto Ricans: Born in the U.S.A* (1991) online edition (http://www.questia.com/read/43049095?title=Puerto Ricans: Born in the U.S.A)
- Samtur, Stephen M. and Martin A. Jackson. *The Bronx: Lost, Found, and Remembered, 1935–1975* (1999) online review (http://www.h-net.org/reviews/showrev.php?id=2982), nostalgia
- Twomey, Bill and Casey, Thomas *Images of America Series: Northwest Bronx* (2011)
- Twomey, Bill and McNamara, John. *Throggs Neck Memories* (1993)
- Twomey, Bill and McNamara, John. *Images of America Series: Throggs Neck-Pelham Bay* (1998)
- Twomey, Bill and Moussot, Peter. *Throggs Neck* (1983), pictorial
- Twomey, Bill. *Images of America Series: East Bronx* (1999)
- Twomey, Bill. *Images of America Series: South Bronx* (2002)
- Twomey, Bill. *The Bronx in Bits and Pieces* (2007)
- Ultan, Lloyd. *The Northern Borough: A History Of The Bronx* (2009), popular general history
- Ultan, Lloyd. *The Bronx in the frontier era: from the beginning to 1696* (1994)
- Ultan, Lloyd. *The Beautiful Bronx (1920–1950)* (1979), heavily illustrated
- Ultan, Lloyd. *The Birth of the Bronx, 1609–1900* (2000), popular
- Ultan, Lloyd. *The Bronx in the Innocent years, 1890–1925* (1985), popular
- Ultan, Lloyd. *The Bronx: It Was Only Yesterday, "The Bronx: It Was Only Yesterday 1935–1965* (1992), heavily illustrated popular history

External links

General links

- Bronx Borough President's Office (http://bronxboropres.nyc.gov/)
- The Bronx Times (http://www.bronx.com/)
- Bronx County, NY Website (http://nybronx.org/)
- Weekly Bronx Report from Inner City Press (http://www.innercitypress.org/bxreport.html)
- Bronx County (http://www.dmoz.org/Regional/North_America/United_States/New_York/Localities/N/New_York_City/The_Bronx/) at the Open Directory Project
- The Bronx River Alliance (http://bronxriver.org/)
- Bronx Council for Environmental Quality (http://www.bceq.org/)
- Throggs Neck Merchant Association (http://www.throggsneckmerchants.com/)
- The Bronx Market (http://www.thebronxmarket.com/)

Places in the Bronx

- The Bronx Museum of the Arts (http://www.bronxmuseum.org/)
- The Bronx Zoo (http://bronxzoo.com/)
- The New York Botanical Garden (http://www.nybg.org/)
- Bronx River Art Center (http://www.bronxriverart.org/)
- Wave Hill: New York Public Garden and Cultural Center (http://www.wavehill.org/home/)
- Walking tour of the Grand Concourse Boulevard-Cross Bronx Expressway area (http://www.pbase.com/patmorgan/the_bronx)
- Walk in the Bronx (http://viewofnewyork.free.fr/View Of New York/Bronx/index.html)
- Yankee Stadium (http://newyork.yankees.mlb.com/nyy/ballpark/index.jsp)

Bronx history

- East Bronx History Forum (http://www.bronxnyc.com)
- Museum of Bronx History (http://www.bronxhistoricalsociety.org/)
- The Bronx County Historical Society (http://www.bronxhistoricalsociety.org)
- The Bronx: A Swedish Connection (http://www.ancestry.com/library/view/ancmag/762.asp)
- Report of the Bronx Parkway Commission, December 31, 1918 (http://books.google.com/books?id=ruhKAAAAMAAJ&pg=PA700), retrieved on July 24, 2008
- Remembrance of Synagogues Past: The Lost Civilization of the Jewish South Bronx (http://www.bronxsynagogues.org) by Seymour Perlin, retrieved on August 10, 2008
- Forgotten New York: Relics of a Rich History in the Everyday Life of New York City (http://www.forgotten-ny.com/)

The Bronx today

- The South Bronx Overall Economic Development Corporation (http://www.sobro.org), retrieved on August 15, 2008
- Melancholy in the Bronx, but Not Because of the Stadium (http://www.nytimes.com/2008/09/20/nyregion/20bronx.html) by David Gonzales, *The New York Times*, published and retrieved on September 19, 2008

125th_Street_(Manhattan)

125th Street is a two-way street that runs east-west in the New York City borough of Manhattan, considered the "Main Street" of Harlem; It is also called Martin Luther King, Jr. Boulevard. Notable buildings along 125th include the West End Theatre (New York) (1902), now home to the La Gree Baptist Church.

West 125th looking from Broadway to the Hudson River

Neighborhoods

The western part of the street runs diagonally between the neighborhoods of Manhattanville and Morningside Heights from the north-west from an interchange with the Henry Hudson Parkway at 130th Street. East of Morningside Avenue it runs east-west through central Harlem to an interchange with F.D.R. Drive by the East River, where it becomes the Manhattan leg of the Triborough Bridge. Many sections of the street have been gentrified and developed with such stores as MAC Cosmetics, Old Navy, H&M, CVS/pharmacy, and Magic Johnson Theaters. The historic Apollo Theater is here. In collaboration with the community, the city has developed a plan for the 125th Street corridor focusing on reinforcing and building upon its strengths as an arts and cultural corridor.[1]

Christmas shopping on 125th Street

West of Convent Avenue, 125th Street was re-routed onto the old Manhattan Avenue. The original 125th Street west of Convent Avenue was swallowed up to make the super-blocks where the low income housing projects now exist. What remains of the original alignment of 125th Street is called La Salle Street, which runs between Amsterdam Avenue and Claremont Avenue.

A proposal to convert the street into a Trans-Harlem Expressway died when funds were diverted from the proposed 125th Street Hudson River bridge.

125th Street between Park Avenue and Madison Avenue

Fault line

A rift in the crust runs along underneath this street and is known as the **125th Street Fault**. The fault line creates a fault valley deep enough to require the IRT Broadway – Seventh Avenue Line to become a trestle bridge between 122nd and 135th Streets. The street in the 18th century was called The Hollow Way.

Public transportation

The following New York City Subway stations are located at 125th Street (west to east):

- 125th Street at Broadway serving the 1 train
- 125th Street at St. Nicholas Avenue serving the A B C D trains
- 125th Street at Lenox Avenue serving the 2 3 trains
- 125th Street at Lexington Avenue serving the 4 5 6 <6> trains

Metro-North Railroad's Harlem – 125th Street commuter station is also on 125th Street, at Park Avenue.

Notable occupants

As of 2011, former president Bill Clinton maintains an office on 125th Street.[2] [3]

55 West 125th where Bill Clinton has his office

- The intersection of 125th and Lexington Avenue is the location where Lou Reed buys heroin on the Velvet Underground's *I'm Waiting for the Man* from their seminal 1967 debut album, *The Velvet Underground & Nico*.[4]

Notes

[1] New York City Economic Development Corporation (NYCEDC) 125th Street project (http://www.nycedc.com/Web/AboutUs/OurProjects/CurrentProjects/125thStreet.htm)

[2] http://www.dnainfo.com/20101108/harlem/bill-clinton-renews-lease-on-harlem-office-space

[3] Leonard, Tom. "Harlem's identity under threat from developers" (http://www.telegraph.co.uk/news/main.jhtml?xml=/news/2008/03/11/wharelm111.xml), *The Telegraph*, March 16, 2008. Accessed April 8, 2008. "Harlem has seen rapid gentrification since the late 1990s and 125th Street - included last year in a list of America's 10 greatest streets - is now home to Bill Clinton's office."

[4] The Velvet Underground, "I'm Waiting for the Man", *The Velvet Underground & Nico*, Verve, 1967.

List_of_New_York_City_Subway_lines

Queens Boulevard viaduct of the IRT Flushing Line

The New York City Subway is a heavy-rail public transit system serving four of the five boroughs of New York City. The present New York City Subway system inherited the systems of the Interborough Rapid Transit Company (IRT), Brooklyn-Manhattan Transit Corporation (BMT), and the Independent Subway System (IND). New York City has owned the IND since its inception; the BMT and IRT were taken over by the city in 1940. The former IRT system is now known as the A Division, while the B Division is the combined former BMT and IND systems.

Primary Trunk line	Color[1] [2]	Pantone [3]	Service bullets
IND Sixth Avenue Line	Bright orange	PMS 165	B D F M
IND Eighth Avenue Line	Vivid blue	PMS 286	A C E
BMT Broadway Line	Sunflower yellow	PMS 116	N Q R
IRT Broadway – Seventh Avenue Line	Tomato red	PMS 185	1 2
BMT Canarsie Line	Light slate gray	50% black	L
IND Crosstown Line	Lime green	PMS 376	G
IRT Flushing Line	Raspberry	PMS Purple	7
IRT Lexington Avenue Line	Apple green	PMS 355	4 5 6

BMT Nassau Street Line	Terra cotta brown	PMS 154	J
Shuttles	Dark slate gray	70% black	S

In the nomenclature of the Subway, the terms "line" and "service" are not interchangeable with each other. While in popular usage the word "line" is often used synonymously with "service" (even sometimes on the website of the MTA[4]), this list will use the formal usage of the term "line."

A *line* is the physical structure and tracks that trains run over. Each section of the system is assigned a unique line name that is paired with its original division (IRT, BMT or IND). For example, the line under Eighth Avenue is the IND Eighth Avenue Line. Some lines have changed names (and even divisions), but this happens relatively infrequently.

By contrast, a *service* refers to the route that a train takes across the various lines. A service can operate along several lines and even along different divisions. For example, the R operates along the IND Queens Boulevard Line as well as the BMT Broadway Line and the BMT Fourth Avenue Line.

Each service is also assigned a *color*. Since 1979, each service's color corresponds to the line it primarily uses in Midtown Manhattan—defined as the *trunk line*—with these exceptions: the IND Crosstown Line, which doesn't carry services to Manhattan, is colored lime green; and all shuttles are colored dark slate gray.[5] The list of trunk lines and colors is shown in the table on the right.

Line listing

There are currently 34 rail lines in service, one (the BMT 63rd Street Line) not in revenue service, and one (the Second Avenue Subway) is under construction. The Archer Avenue Line and the 63rd Street Line are classified as two separate lines each due to their structure: both lines are able to serve two divisions (the BMT and the IND) on distinct sections of track.

In the list below, lines with colors next to them indicate trunk lines, which determine the colors that are used for services' route bullets and diamonds. The opening date refers to the opening of the first section of track for the line. In the "division" column, the current division is followed by the original division in parentheses.

Division	Line	Borough(s)	Service(s)	Opened	Structure
B (IND)	Second Avenue Line	Manhattan	N/A	under construction	underground
B (BMT)	Fourth Avenue Line	Brooklyn	D ● N ● R ●	June 22, 1915[6]	underground
B (IND)	Sixth Avenue Line	Manhattan Brooklyn	B ○ D ● F ● M ○	January 1936[7]	underground
B (IND)	Eighth Avenue Line	Manhattan Brooklyn	A ● B ○ C ◎ D ● E ●	September 10, 1932[8]	underground
A (IRT)	42nd Street Shuttle	Manhattan	S ◎	October 27, 1904[9]	underground[a]
B (BMT)	63rd Street Line	Manhattan	no regular service	October 29, 1989[8]	underground

B (IND)	63rd Street Line	Manhattan Queens	F ●	October 29, 1989[8]	underground
B (BMT)	Archer Avenue Line	Queens	J ● Z ◆	December 11, 1988[6] [8]	underground
B (IND)	Archer Avenue Line	Queens	E ●	December 11, 1988[8]	underground
B (BMT)	Astoria Line	Queens	N ● Q ○	April 21, 1917	elevated
B (BMT)	Brighton Line	Brooklyn	B ○ Q ●	July 2, 1878[10]	underground, open cut, embankment, elevated
B (BMT)	Broadway Line	Manhattan	N ● Q ● R ◎	September 4, 1917[6]	underground
A (IRT)	Broadway – Seventh Avenue Line	Bronx Manhattan Brooklyn	1 ● 2 ● 3 ●	October 27, 1904[9]	elevated[a], embankment, underground
B (BMT)	Canarsie Line	Manhattan Brooklyn	L ●	October 21, 1865	underground, elevated, at-grade
B (IND)	Concourse Line	Bronx Manhattan	B ○ D ●	July 1, 1933[8]	underground
B (IND)	Crosstown Line	Brooklyn Queens	G ●	August 19, 1933[8]	underground
B (IND)	Culver Line	Brooklyn	F ● G ●	March 16, 1919[6]	underground, elevated[c]
A (IRT)	Dyre Avenue Line	Bronx	5 ●	May 15, 1941	elevated[b], embankment, open-cut, underground
A (IRT)	Eastern Parkway Line	Brooklyn	2 ● 3 ◎ 4 ● 5 ○	January 9, 1908[9]	underground
A (IRT)	Flushing Line	Manhattan Queens	7 ● <7> ◆	June 22, 1915[9]	underground, elevated
B (BMT)	Franklin Avenue Line	Brooklyn	S ●	July 2, 1878[10]	elevated, embankment, open cut
B (IND)	Fulton Street Line	Brooklyn Queens	A ● C ◎	April 9, 1936[8]	underground, elevated[d]
B (BMT)	Jamaica Line	Brooklyn Queens	J ● M ● Z ◆	February 2, 1885[6]	elevated
A (IRT)	Jerome Avenue Line	Bronx	4 ● 5 ◎	June 12, 1917[9]	elevated, underground
A (IRT)	Lenox Avenue Line	Manhattan	2 ● 3 ●	November 23, 1904[9]	at-grade, underground
A (IRT)	Lexington Avenue Line	Manhattan	4 ● 5 ◎ 6 ● <6> ◆	October 27, 1904[9]	underground[a]

B (BMT)	Myrtle Avenue Line	Brooklyn Queens	M ●	December 19, 1889[6]	elevated, embankment, at-grade
B (BMT)	Nassau Street Line	Manhattan	J ● M ○ Z ◆	September 16, 1908[6]	underground
A (IRT)	New Lots Line	Brooklyn	2 ◆ 3 ◎ 4 ◐ 5 ◆	November 23, 1920[9]	elevated
A (IRT)	Nostrand Avenue Line	Brooklyn	2 ● 5 ○	August 23, 1920[9]	underground
A (IRT)	Pelham Line	Bronx	6 ● <6> ◆	August 1, 1918[9]	underground, elevated
B (IND)	Queens Boulevard Line	Manhattan Queens	E ● F ● M ○ R ◎	August 19, 1933[8]	underground
B (IND)	Rockaway Line	Queens	A ● S ●	June 28, 1956	at-grade, embankment, elevated[e]
B (BMT)	Sea Beach Line	Brooklyn	N ●	June 22, 1915[6]	open cut, elevated
B (BMT)	West End Line	Brooklyn	D ●	June 24, 1916[6]	open-cut, elevated
A (IRT)	White Plains Road Line	Bronx	2 ● 5 ●	July 10, 1905[9]	elevated, underground

Connections

The following list shows the connections between the different divisions of the New York City Subway.

Purpose-Built

These connections can be used by trains in revenue service:

- 60th Street Tunnel Connection (Queens) (BMT/IND) (R train)
- 63rd Street Lines (BMT 63rd St Line / IND 63rd St Line) (Manhattan) (BMT/IND)
- Coney Island – Stillwell Avenue station (BMT/IND) (originally all BMT) (D F N Q trains)
- Chrystie Street Connection (Manhattan) (BMT/IND) via the Manhattan Bridge (B D trains) and via the Williamsburg Bridge (M train)

This connection is not for revenue service due to the differing widths of the trains:

- Astoria Line / Flushing Line (Queens) (BMT/IRT)

Yards

These connections are located within the subway's rail yards and are not intended for revenue service.

- 207th Street Yard (Manhattan) (IND/IRT); trains from the IND Eighth Avenue Line and IRT Broadway – Seventh Avenue Line
- Coney Island Complex (Brooklyn) (BMT/IND); trains from the BMT Brighton Line, IND Culver Line, BMT Sea Beach Line and BMT West End Line
- Concourse Yard (Bronx) (IND/IRT); trains from the IND Concourse Line and IRT Jerome Avenue Line
- Linden Yard (Brooklyn) (BMT/IRT/LIRR); trains from the BMT Canarsie Line and IRT New Lots Line

Other

- South Brooklyn Railway / BMT West End Line (Brooklyn) (BMT/Mainline RR)

Defunct lines

The following New York City Subway lines are either entirely defunct or have major portions no longer in service. Defunct spur lines with one station, such as the South Ferry loops, are not included in this list, nor are surface transit lines.

Division	Line	Borough(s)	Opened	Closed	Structure	Status
A (IRT)	Second Avenue Line	Manhattan	March 1, 1880	June 13, 1942	elevated	Entirely demolished
A (IRT)	Third Avenue Line	Manhattan Bronx	August 26, 1878	April 28, 1973	elevated	ended service to Manhattan on May 12, 1955 Entirely demolished
B (BMT)	Third Avenue Line	Brooklyn			elevated	Entirely demolished
B (BMT)	Fifth Avenue Line	Brooklyn			elevated	Entirely demolished
A (IRT)	Sixth Avenue Line	Manhattan	1878	December 4, 1938	elevated	Entirely demolished
A (IRT)	Ninth Avenue Line	Manhattan Bronx	July 1, 1868	August 31, 1958	elevated	ended service to Manhattan on June 11, 1940 Entirely demolished; except for the two underground stations at Sedgwick Avenue and Anderson – Jerome Avenue in the Bronx
B (BMT)	Canarsie Line	Brooklyn	19th Century	November 21, 1942	at-grade	Line from Canarsie – Rockaway Parkway to Canarsie Pier replaced with streetcars, then abandoned and replaced with buses.
B (BMT)	Culver Line	Brooklyn			elevated	Line from Ditmas Avenue to Coney Island – Stillwell Avenue taken over by the IND in 1954;[c] remainder of line to Ninth Avenue abandoned in 1975 and demolished in 1985.
B (BMT)	Fulton Street Line	Brooklyn Queens			elevated	Line west of Hudson Street – 80th Street demolished;[d] remainder of line taken over by the IND in 1956.
A (IRT)	IRT trunk line	Manhattan	October 27, 1904	July 17, 1918	underground	Separated into the Broadway – Seventh Avenue, Lexington Avenue and 42nd Street Shuttle lines[a]
B (BMT)	Lexington Avenue Line	Brooklyn			elevated	Line west of Gates Avenue demolished

B (BMT)	Jamaica Line	Brooklyn Queens			elevated	Line west of Marcy Avenue demolished Line east of 121st Street demolished
B (BMT)	Myrtle Avenue Line	Brooklyn Queens	1888-1924	1944-1969	elevated	Line west of Broadway demolished
B (IND)	World's Fair Line	Queens	1939	1940	at-grade	Entirely demolished

See also

- List of New York City Subway services
- New York City Subway nomenclature
- Staten Island Railway

Notes

- [a] The route of the original IRT line, the first underground New York City rapid transit line, began at City Hall in the south, followed the IRT Lexington Avenue Line to 33rd Street, turned west on 42nd Street to Grand Central, followed the IRT 42nd Street Shuttle to Times Square, turned north on Broadway to 50th Street, followed the IRT Broadway – Seventh Avenue Line and terminated at 145th Street.
- [b] The IRT Dyre Avenue Line was formed from the purchase of abandoned New York, Westchester and Boston Railway trackage in 1941.
- [c] The formation of the IND Culver Line occurred when the IND "recaptured" (e.g. New York City took over lines it had built for private companies) the trackage rights of the BMT Culver Line in 1954, connecting Ditmas Avenue to Church Avenue, the terminus of the IND Brooklyn Line.
- [d] For the extension of the IND Fulton Street Line in 1956, the IND "recaptured" the trackage rights of the BMT Fulton Street Line, connecting Hudson Street–80th Street to Grant Avenue.
- [e] The IND Rockaway Line was formed from the purchase of some trackage from the Long Island Rail Road's Rockaway Park Branch and Far Rockaway Branch in 1955.

References

[1] Official paint monikers since the colors were fixed in 1979: Grynbaum, Michael (May 10, 2010). "Take the Tomato 2 Stops to the Sunflower" (http://cityroom.blogs.nytimes.com/2010/05/10/take-the-tomato-to-the-sunflower/). *New York Times, City Room Blog*. . Retrieved May 11, 2010.

[2] Official MTA video mentions "lime green" for the G line. "Subway Colors and Names" (http://www.youtube.com/watch?v=yZ83UhBJFP0). *MTA Info*. July 15, 2010. . Retrieved August 5, 2010.

[3] MTA Developer Resources Download (http://www.mta.info/developers/download.html), CSV file (http://www.mta.info/developers/data/colors.csv)

[4] Subway Line Information (http://mta.info/nyct/service/index.html)

[5] Hogarty, Dave (August 3, 2007). "Michael Hertz, Designer of the NYC Subway Map" (http://gothamist.com/2007/08/03/michael_hertz_d.php). *Gothamist*. . Retrieved July 4, 2009.

[6] "New York City Subway BMT Division Timeline" (http://www.nycsubway.org/articles/bmttime.html). www.nycsubway.org. . Retrieved 2009-05-07.

[7] "New York City Subway IND 6th Ave Line" (http://www.nycsubway.org/lines/6thave.html). www.nycsubway.org. . Retrieved 2009-05-07.

[8] "New York City Subway IND Division Timeline" (http://www.nycsubway.org/articles/indtime.html). www.nycsubway.org. . Retrieved 2009-05-07.

[9] "New York City Subway IRT Division Timeline" (http://www.nycsubway.org/articles/irttime.html). www.nycsubway.org. . Retrieved 2009-05-07.

[10] "New York City Subway Brighton Beach Line" (http://www.nycsubway.org/lines/brighton.html). www.nycsubway.org. . Retrieved 2009-05-07.

Unused_New_York_City_Subway_service_labels

The New York City Subway currently uses various letters and numbers to designate the routes that trains use over the differing lines in the system. Along with the color corresponding to the route's trunk line, these form a unique identifier for the route, easing navigation through the complex system. Several service labels have either been phased out or never been used. This list covers the labels not used as of January 2011.

A Division numbers

The A Division uses single-digit numbers for each route. 1 through 7 are in use.

- **0** is used as the internal designation of the 42nd Street Shuttle.
- **8** has been used in the past for various routes, most recently the Bronx portion of the Third Avenue El, ended in 1973.
- **9** was used for skip-stop service on the IRT Broadway – Seventh Avenue Line from 1989 to 2005.

An R62A car in Corona Yard displays a **12** sign.

Two-digit numbers have never been used by the A Division, but have been seen on the current rollsigns of some trains, paired with colors used with other services. It is likely that these were assigned arbitrarily, for use if the MTA changed the additional rush hour express service designators from a "diamond" version of the regular number to a separate number.

B Division letters

Trains of the B Division use single letters of the English alphabet. These service letters are unused as of January 2011, but some have been used or proposed for services at various points in time:

- **H** was last used publicly for the Rockaway Park Shuttle in 1993, but is still used as its internal-only designator by NYC Transit. Since November 2012 the letter has been also used for a fare-free shuttle service in the Rockaways after damage caused by Hurricane Sandy rendered normal A train service in the area inoperable.[1] [2] [3]
- **I** has never been used due to its visual similarity to the number 1.[4]
- **K** has been used for multiple routes in the past, most recently as an Eighth Avenue Line train (discontinued in 1988).
- **O** has never been used due to its visual similarity to the number 0.[4]
- **P** (short for Penn Station) was planned for a nonstop between Sutphin Boulevard – Archer Ave (at the LIRR's Jamaica station) and 34th Street – Penn Station, continuing locally to 168th Street, via the BMT Archer Avenue Line, BMT Jamaica Line, Chrystie Street Connection, IND Sixth Avenue Line and IND Eighth Avenue Line (switching to the latter at West Fourth Street) in the 1990s. This would have been used during a threatened Amtrak strike that would have prevented LIRR trains from entering Penn Station.[5]
- **T** was used for BMT West End Line services, but was quickly replaced by the expanded **B** service after services were reorganized following the opening of the Chrystie Street Connection. T is the planned label for the new Second Avenue Line.

- **U** has never been used, as it sounds like the word "you".[4]
- **V** was introduced in 2001 for weekday local service on the IND Queens Boulevard Line and IND Sixth Avenue Line via 53rd Street, after the **F** was shifted to the newly-opened IND 63rd Street Line. V service was discontinued in 2010 due to budgetary concerns, replaced almost entirely by a rerouted **M** train.[6]
- **W** was introduced in 2001 for BMT West End Line—BMT Broadway Line service via the south side of the Manhattan Bridge; reconstruction of the bridge's north tracks prevented the usual **B** service from running to the West End Line. After work was completed, the W became a weekday local service on the BMT Broadway Line and BMT Astoria Line, supplementing the **N** in Queens and the **R** in Manhattan. This service was discontinued in 2010 due to budgetary concerns, replaced by the N and **Q** in Queens and the N and R in Manhattan.[6]
- **X** is used as a placeholder for subway lines under construction.[4]
- **Y** has never been used, as it sounds like the word "why".[4]
- The JFK Express bullet, featuring an airplane inside a blue circle, was used from 1978 till 1990.

Prior to 1985, the B Division used two-letter combinations to indicate differing variations of similar services, but these were phased out in favor of single-letter designators. These former service names are covered in their corresponding current letters.

References

[1] MTA's tweet on partially restoring the Rockaway shuttle (https://twitter.com/MTAInsider/status/270625005337993216)

[2] Hurricane Sandy Recovery Service As of November 20 (http://alert.mta.info/sites/default/files/pdf/hurricane_recovery_map_bw_Nov_20_rock_shtl.pdf)

[3] Beginning Tuesday, Fare-Free H Shuttle Replaces Portion of A Line in Rockaways (http://mta.info/news/stories/?story=878)

[4] V and W Trains join a long list of routes that have bowed out of the Subways (http://amny.com/urbanite-1.812039/v-and-w-trains-join-a-long-list-of-routes-that-have-bowed-out-of-the-subways-1.2028944), AM New York. Accessed 2010-08-21

[5] Ritea, Steve (2008-01-10). "LIRR preps for strike that might shut Penn Station" (http://www.newsday.com/long-island/lirr-preps-for-strike-that-might-shut-penn-station-1.880134). *Newsday*. . Retrieved 2009-09-26.

[6] "Major Subway Changes Set for Monday" (http://www.mta.info/mta/news/releases/?en=100624-NYCT90). Metropolitan Transportation Authority. June 24, 2010. . Retrieved December 24, 2010.

See also

New York City Subway nomenclature

Second_Avenue_(Manhattan)

Second Avenue

Second Avenue is an avenue on the East Side of the New York City borough of Manhattan extending from Houston Street at its south end to the Harlem River Drive at 128th Street at its north end. A one-way street, vehicular traffic runs only downtown. A bicycle lane in the left hand portion from 55th to 34th Street closes a gap in the Manhattan Waterfront Greenway. There is also a bike lane from 34th Street all the way down to Houston Street. Second Avenue passes through a number of Manhattan neighborhoods including (from south to north) the Lower East Side, East Village, Gramercy Park, Murray Hill, Upper East Side, Yorkville and Spanish Harlem.

History

Downtown Second Avenue in the Lower East Side was the home to many Yiddish theatre productions during the early part of the 20th century, and Second Avenue came to be known as "Yiddish Broadway" or the "Jewish Rialto". Although the theaters are gone, many traces of Jewish immigrant culture remain, such as kosher delicatessens and bakeries, and the famous Second Avenue Deli (which closed in 2006, later reopening on East 33rd Street and Third Avenue).

Second Avenue facing north from 42nd Street in 1861

The Second Avenue Elevated train line ran above Second Avenue the full length of the avenue north of 23rd Street, and stood from 1880 until service was ended on June 13, 1942. South of Second Avenue, it ran on First Avenue and then Allen and Division Street.[1] The elevated trains were noisy and often dirty (in the 19th century they were pulled by soot-spewing steam locomotives). This depressed land values along Second Avenue during the late 19th and early 20th centuries. Partially because of the presence of the El, most buildings constructed during this era were working class tenements. The line was finally torn down in 1942 because it was deteriorated and obsolete, and the cost of World War II made upkeep impossible.[2] Second Avenue maintains its modest architectural character today, despite running through a number of high income areas.

Second Avenue has carried one-way traffic since June 4, 1951.[3]

Public transportation

Bus service

The M15 local serves the entirety of Second Avenue. In addition, the M15 Select Bus Service provides a faster way of bus transportation. Additionally, the M34A Select Bus Service runs along Second Avenue between East 34th Street and East 23rd Street en route to Waterside Plaza.

Subway

A Second Avenue Subway line has been planned since 1919. A subway line under Second Avenue would relieve congestion on the IRT Lexington Avenue Line (4 5 6 <6>), currently the only line running up the East Side of Manhattan. A few short sections of the line have been completed over the years, with some serving other subway lines (the Grand Street Station is one example), and others simply sitting vacant underground (such as the unused upper level at the Second Avenue station). Portions have been leased from time to time by New York Telephone to house equipment serving the company's principal north-south communication lines which run under the Avenue. Construction was begun again on April 12, 2007, on Phase 1 of a subway line that will eventually extend from 125th Street to the Financial District. Phase 1 connects the currently unused BMT 63rd Street Line with the new line north to stations at 72nd, 86th and 96th Streets.

References

[1] Staff. "Second Avenue 'El' Coming to a Stop" (http://pqasb.pqarchiver.com/csmonitor_historic/access/278313072.html?dids=278313072:278313072&FMT=ABS&FMTS=ABS:AI&date=Jun+13,+1942&author=By+a+Staff+Correspondent+of+The+Christian+Science+Monitor&pub=Christian+Science+Monitor&desc=Second+Avenue+'El'+Coming+to+a+Stop&pqatl=google), *The Christian Science Monitor*, June 13, 1942. Accessed October 12, 2008.

[2] http://www.nycsubway.org/articles/2ndave-route132c.html

[3] Ingraham, Joseph (5 June 1951). "Autos Speeded 15% on 1st And 2nd Aves" (http://select.nytimes.com/gst/abstract.html?res=F40C13F8395D147B93C7A9178DD85F458585F9). *The New York Times*. . Retrieved 28 August 2012.

External links

- New York Songlines: Second Avenue (http://www.nysonglines.com/2av.htm), a virtual walking tour

List_of_New_York_City_Subway_inter-division_conne

The Grand Street station was built as part of the Chrystie Street Connection

The New York City Subway was formed from three different systems, the IRT, BMT and IND. For operational purposes, the IRT is *A Division* and the BMT and IND make *B Division*; however, common usage calls the three systems divisions. Over the years, various connections have been built to allow trains of one division to operate on the lines of another.

For transfers, see List of New York City Subway transfer stations.

Revenue connections

Because the IRT system uses narrower tunnels and cars than the BMT and IND, the only possible inter-division revenue connections are between BMT and IND lines. The following connections have been built:

- October 30, 1954: IND Culver Line (F train) (originally called the Smith Street Subway) extended over BMT Culver Line; original routing disconnected and became the Culver Shuttle until 1975
- December 1, 1955: 60th Street Tunnel Connection (R train) between the IND Queens Boulevard Line and the BMT 60th Street Tunnel
- April 29, 1956: IND Fulton Street Line (A train) joined to BMT Liberty Avenue Elevated (part of the Fulton Street Elevated, but disconnected from the BMT system and so is only used for IND service)
- November 26, 1967: Chrystie Street Connection between the IND Sixth Avenue Line and the BMT Nassau Street Line (via the Williamsburg Bridge – M train) and BMT Fourth Avenue Line (via the Manhattan Bridge – B D trains)
- October 29, 1989: connecting tracks between IND 63rd Street Line and BMT 63rd Street Line (not generally used for revenue service unless a re-route of the R train is required north of 57th Street on the BMT Broadway Line)

Non-revenue connections

These connections mainly take place in yards:

- Crossover tracks (with third-rails) between northbound IRT Flushing Line and BMT Astoria Line at Queensboro Plaza (the Flushing Line's only connection to the rest of the system, originates from when both lines north of Queensboro Plaza were dual-operated)
- IRT New Lots Line and BMT Canarsie Line both have a single non-electrified track into the Linden Shops. The track leaves the New Lots Line at Junius Street and joins the track from Canarsie line south of Livonia Avenue before arriving in the Linden complex
- IRT Broadway – Seventh Avenue Line and IND Eighth Avenue Line both have tracks to 207th Street Yard
- IRT Jerome Avenue Line and IND Concourse Line both have tracks to Concourse Yard
- Since October 30, 1954, the IND Culver Line has gone into Coney Island – Stillwell Avenue, where it connects to the BMT Brighton Line, BMT Sea Beach Line and BMT West End Line. For a short time in the late 1960s the NX service passed through Coney Island – Stillwell Avenue to and from the Sea Beach and Brighton Lines, but no service has passed through from the Culver Line.

New_York_City_Subway_yards

The following rail yards serve the New York City Subway:

A Division Yards

137th Street Yard

The **137th Street Yard** is an underground rail yard located just north of 137th Street – City College on the IRT Broadway – Seventh Avenue Line, hence the yard's name. It is composed of five tracks surrounding the three mainline tracks. Three tracks are to the west of the line (downtown side) and two are to the east (uptown side). The yard is used to store trains for the **1** service. 40°49′22″N 73°57′11″W

239th Street Yard

239th Street Yard aerial view

239th Street Yard east side

R110As and R62As in the 239th Street Yard

The **239th Street Yard** is the northern most rail yard located at 4570 Furman Avenue in Wakefield and is home to the R142s assigned to the **2** train. There is considerable fleet interchange with the R142s at East 180th Street Yard used for **5** service. The car wash at 239th Street is used for both services. A wheel truing machine was installed here to minimize damage to rail cars and tracks caused by flat wheels. This shop was also used as a facility to retrofit all R26s, R28s, R29s, R33s (except single unit Worlds Fair cars) and R36s (both Mainline and Worlds Fair types) married pairs IRT type cars with the installation of new Stone Safety 10 ton Air Conditioning systems and units between 1976 and 1981. Also during this period all cars assigned to the **2** train was inspected and maintained at the East 180th Street Shops shared with the **5** trains. It reopened as an inspection and maintenance facility for the **2** train in 1982.

The yard consists of a seven-track inspection shop and 38 layup tracks. The layup tracks are arranged on two levels; the only other yard in the system to share this trait is the East New York Yard. Access to the **239th Street Yard** is located between Wakefield – 241st Street and Nereid Avenue on the IRT White Plains Road Line. 40°54′03″N 73°50′58″W

240th Street Yard

The **240th Street Yard**, also known as **Van Cortlandt** or **VC Yard**, is located at 5911 Broadway in the Bronx, serving the IRT Broadway – Seventh Avenue Line. The yard consists of six inspection tracks in the shop and 15 additional layup tracks. The yard is presently home to the R62As assigned to the **1** train. The shop was built in 1906 to support the original IRT subway. The yard and shops are entirely on an elevated structure. 40°53′18″N 73°54′05″W

240th Street Yard layup tracks

Corona Yard

Corona Yard serves the IRT Flushing Line (**7** **<7>** trains). It is located in Flushing Meadows–Corona Park near Citi Field, the National Tennis Center, and the site of the 1939 and 1964 World's Fairs.

Corona Yard and shop

Corona Yard opened in 1928 and has seen various models of cars, including Steinway Low-Vs, World's Fair Lo-Vs, BMT Qs, R12s, R14s, R15s, some R17s from 1961 thru early 1965 to enable operation of 11 car trains, and were often mixed with R15s. Some R26s & R28s & R29s were temporarily assigned from 1978 to 1985 due to its assigned fleet of World's Fair R33s, & World's Fair R36s being air conditioned (the World's Fair R33s did not receive air conditioning), and overhauled. And of course a new fleet of World's Fair R33s, & World's Fair R36s in 1963 and 1964. Also R33s from 1989-1996 & 2002-2003 to temporarily fill in due to light overhaul work on all WFR36 cars done between 1990 and 1996, and also as a fill in due to retirements of many Westinghouse equipped WFR36's during 2002 and 2003. and the currently assigned R62As. It also contains the Casey Stengel Bus Depot.

The Corona Yard at night

On August 16, 2006, the original 1928 shop building was demolished, and was replaced by a new, modern shop. 40°45′10″N 73°50′46″W

East 180th Street Yard

The **East 180th Street Yard** is situated at 1145 East 180th Street in The Bronx, just east of the Bronx Zoo. The yard consists of eight storage tracks and an adjacent 12-track shop building with a connection to the nearby 19-track Unionport Yard, which lies to the northeast of East 180th Street Yard.[1] The yard is currently home of the R142s for the **5** train. There is also some equipment interchanged with the **2** train. 40°50′33″N 73°52′22″W

The East 180th Street Yard and Maintenance–Repair Shop; viewed from East 180th Street station

Jerome Avenue Yard

The **Jerome Yard**, also known as **Mosholu Yard**, is located at 3191 Jerome Avenue in The Bronx. This yard is home to all cars assigned to the **4** train. It is one of the three yards in the system to be under a housing complex (Pitkin Yard and Lenox Yard are the others). Rail access to the yard is by a pair of tracks that branch off of the elevated IRT Jerome Avenue Line just north of Bedford Park Boulevard – Lehman College station. The riveted steel pylons that support the elevated branch tracks give way to stone pylons just north of 205th Street before they enter the yard. The yard is surrounded by a wall and covered by a parking deck used by residents of the Tracey Towers housing complex. 40°52′44″N 73°53′16″W

Lenox Yard

Lenox Yard is located near 148th Street and Lenox Avenue in Harlem, New York. It has no maintenance facility and is only used as storage of trains that operate on the **3** service. It was the first overhaul shop for the IRT and opened with the subway in 1904. Two tracks were taken from the yard in 1968 for Harlem – 148th Street station, which is the northern terminal for the **3** train. 40°49′22″N 73°56′07″W

Lenox Yard in 1902

Livonia Yard

The **Livonia Yard** is in East New York, Brooklyn on an entirely elevated structure at the east end of the IRT New Lots Line. One of the smaller yards in the system, it is home to the R62s on the **3** train and the ten R62As used for the 42nd Street Shuttle.

The yard consists of four inspection tracks in the shop and thirteen layup tracks. Many **3** trains are stored in the Lenox Yard in Upper Manhattan, as Livonia is not very large, and a notable amount of space within Livonia is used for the storage of **2**, **4** and **5** trains. 40°39′51″N 73°52′45″W

Northwest end of Livonia Yard

Unionport Yard

Unionport Yard is associated with the nearby East 180th Street Yard, used primarily as a lay-up facility for the **2** and **5** trains.[2] [1] It is named after Unionport Road which lies just east of the yard. There are no shop or wash facilities at this yard, which was expanded in the 1990s from five tracks to its present 19. All but one track ends in bumper block. The newly-expanded yard was fully operational in 1997. The yard connects to the IRT White Plains Road Line to the south and the IRT Dyre Avenue Line to the north. 40°50′42″N 73°52′12″W

West Farms Yard (demolished)

The **West Farms Yard** was an elevated rail yard that was demolished in 1952 along with the redundant 180th Street – Bronx Park station.[3] It formerly served trains of the 2 and 5 services.

Westchester Yard

The **Westchester Yard**, also known as the **Pelham Yard**, is located in the Bronx. It has 45 layup tracks and is home to the **6 <6>** trains as well as Maintenance of Way diesel trains for both the A Division and B Division. It is connected to the IRT Pelham Line in both directions between Westchester Square – East Tremont Avenue and Middletown Road stations.

There is a four-track inspection shed for electric trains and a two-track diesel repair shop. Pelham Yard also has a car wash used by the entire A Division. 40°50′38″N 73°50′31″W

B Division Yards

174th Street Yard

The **174th Street Yard** is an underground rail yard on the IND Eighth Avenue Line, used to store trains for the **C** service. It is composed of five tracks to the east of the two mainline passenger service tracks. The yard is located six blocks north of 168th Street and adjacent to 175th Street. The inner tracks at 168th Street lead towards the yard and are used by terminating **C** trains. This yard can hold only three full-length trains despite having five tracks. 40°50′45″N 73°56′23″W

207th Street Yard

The **207th Street Yard** is located in upper Manhattan on the Harlem River north of the University Heights Bridge. It serves as the home yard for the R32s assigned to the C train.

View of 207th Street Yard from the southwest - note the Museum cars.

The 207th Street Shop is one of two heavy overhaul shops in the New York City Subway system (the other being the Coney Island Yard in Brooklyn) and provides overhaul and rebuilding of most A Division cars as well as some B Division rolling stock.

The yard also stores cars that are being retired or awaiting scrapping and restores cars designated for the New York Transit Museum.

South of the yard, connecting tracks lead to the IND Eighth Avenue Line. A separate connecting track and flyover leads to the IRT Broadway – Seventh Avenue Line. 40°52′0″N 73°55′0″W

View of 207th Street Yard from a nearby building

Canarsie Yard

The **Canarsie Yard** (also known as **AY Yard** from its telegraphy letters) is located on the south end of the BMT Canarsie Line adjacent to Canarsie – Rockaway Parkway. It is the primary layup yard for the **L** train and hosts the only car wash for the entire BMT Eastern Division, washing trains for the **J**, **L**, **M** and **Z** services. Signals were installed in 2003 in conjunction with the BMT Canarsie Line automation project. This greatly increases the safety of train movements in the yard, as the **L** train has some of the fastest growing ridership in the system. 40°38′48″N 73°54′05″W

Church Avenue Yard

The **Church Avenue Yard** is an underground rail yard on the IND Culver Line, used to store trains for the **G** service. It is composed of four tracks directly under the four main line tracks above. There are two ramps between each local and express track south of Church Avenue station for access. Each track can hold one full-length train between the bumper blocks and the crossovers. 40°38′31″N 73°58′45″W

Coney Island Complex

Coney Island Yard street entrance on Shell Road

The **Coney Island Complex** is the largest rapid transit yard in the world. Located in Brooklyn, New York, it covers 75 acres (m^2) and operates 24/7.

Coney Island Complex

The complex was built in 1926 on former marshlands that, along with Coney Island Creek, formerly separated Coney Island from the main body of Brooklyn. Much of this land had originally been proposed for use as a ship canal and port facility.

Regular scheduled maintenance is performed here for a fleet of nearly 800 cars serving the **B**, **G**, **N**, **Q** and **Franklin Avenue Shuttle** trains. The shop facility along with 207th St. does heavy maintenance and overhaul for every one of the approximately 6,000 cars in the subway system, including the Staten Island Railway.

In addition to heavy maintenance facilities and track facilities for cars undergoing maintenance and overhaul, the complex includes three related railroad storage yards. The main yard facility, known as "Coney Island Yard," includes direct connections to the adjacent BMT Sea Beach Line (N train) and a two-track elevated structure to the BMT

West End Line (D train). The main yard also serves trains on the BMT Brighton Line (B Q trains) via tracks C & D (aka 3 & 4) of Coney Island – Stillwell Avenue station. The adjacent but separate Culver (or "City") Yard connects to the IND Culver Line (F train) at the eastern border of the yard complex. Another yard, "Stillwell Yard," used mainly for off-peak train storage, is located across the Sea Beach Line from the main yard complex in a "V" between the divergent Sea Beach and West End Lines.

Inside view into a workshop

In addition to the maintenance shop and yards, there is a Health Center (gym) for Transit Authority employees and a firing range for the New York City Police Department (NYPD). The range was originally built for the New York City Transit Police Department, which was merged with the NYPD in 1995.

The Coney Island Yard Electric Motor Repair Shop and Coney Island Yard Gatehouse were listed on the National Register of Historic Places in 2006.[4] 40°35′10″N 73°58′40″W

Concourse Yard

Concourse Yard Entry Buildings	
U.S. National Register of Historic Places	

Location:	W. 205th St., bet. Jerome and Paul Aves., Bronx, New York
Coordinates:	40°52′34″N 73°53′21″W
Area:	less than one acre
Built:	1933
Architect:	Ridgeway, Robert
Architectural style:	Art Deco
Governing body:	State
MPS:	New York City Subway System MPS
NRHP Reference#:	06000014 [4]
Concourse Yard Substation	
U.S. National Register of Historic Places	
Location:	3119 Jerome Ave., Bronx, New York

NRHP Reference#:	06000013 [4]
Added to NRHP:	February 9, 2006

Concourse Yard from the southeast

The **Concourse Yard** is located in northern Bronx near 205th Street and Jerome Avenue. This yard is home to all cars assigned to the **D** trains and are maintained at this yard, although many trains assigned to the **B** and **4** services are stored here as well. Connecting tracks lead north from the yard to the IND Concourse Line and south to the IRT Jerome Avenue Line. Concourse Yard is spanned across it middle by Bedford Park Boulevard West, and at its northern end by a 205th Street viaduct. The Jerome Yard used by the IRT Jerome Avenue Line lies to the north of 205th Street.

Concourse Yard maintenance building

The **Concourse Yard Entry Buildings** and **Concourse Yard Substation** were listed on the National Register of Historic Places in 2006.[4] The former consists of two, three-story brick buildings with only the top story visible from the street that are built next to one another to form a gateway to the Concourse Yard. They feature ornamental limestone columns and aluminum doors. The buildings are connected by an iron bridge that retains its original Art Deco balustrade.[5] The latter is a one-story brick building measuring 50 feet by 100 feet and featuring a brick parapet with ornamental limestone and aluminum doors.[5] 40°52′28″N 73°53′25″W

East New York Yard

East New York Yard (also known as **DO Yard** from its telegraphy letters) is primarily used to store and repair trains on the J, L, M and Z trains. Subway equipment is inspected and maintained there on a regular basis.[6]

It is located at the junction of the Canarsie and Jamaica Lines near the intersection of Broadway and Jamaica Avenue in East New York, Brooklyn. A separate part of the facility houses the East New York Bus Depot, formerly a trolley depot. The yard is entirely equipped with hand-operated switches. Only the Fresh Pond Yard and 36th–38th Street Yard share this characteristic.

Portions of the yard date back to 1885 and the Lexington Avenue Elevated and the yard predates the rebuilding of nearby Broadway Junction, which was formerly known as **Manhattan Junction** or **East New York Loop**.

The yard and its main lead configuration remained the same before and after the extensive elevated line rebuilding nearby, but additional track and structure was built, so that, at its peak, East New York Yard had direct connections to the Broadway Elevated going west, Jamaica Line going east, Canarsie Line going east, and Fulton Street Elevated both east and west. 40°40′44.3″N 73°54′6.1″W

Fresh Pond Yard

The **Fresh Pond Yard** in Middle Village, Queens is used for storing the R160As that run on the **M** train. It is located between Fresh Pond Road and Middle Village – Metropolitan Avenue on the BMT Myrtle Avenue Line, but is only accessible from the latter station. Trains must first platform there and then reverse into the yard. The yard is entirely equipped with hand-operated switches. Only East New York Yard and 36th–38th Street Yard share this characteristic. 40°42′26″N 73°53′36″W

Jamaica Yard

Southwestern corner of Jamaica Yard

Employee entrance of Jamaica Yard

Jamaica Yard is located in Kew Gardens, Queens at the south end of Flushing Meadows–Corona Park. It is connected to the IND Queens Boulevard Line at a three-way flying junction. The yard is on surface level, and the four-track approach includes a bridge over the Grand Central Parkway, despite the Queens Boulevard Line being underground.[7] The site upon which the yard sits at the head of the valley of the Flushing River was, during the American Revolution, occupied by British troops after the Battle of Long Island.[8]

The yard was built for the Independent Subway System in the 1930s and served as the south end of the World's Fair Railroad in 1939 and 1940. The yard provides carwash, interior cleaning, grease and minor repair services to the R46s and R160s that are assigned to the **E**, **F**, and **R** trains.[9] [10]

The Metropolitan Transportation Authority plans to enlarge the yard.[11] 40°43′12″N 73°49′41″W

Pitkin Yard

The **Pitkin Yard** is located in East New York, Brooklyn. It holds the Pitkin Shops, which maintain the R46s used on the **A** train and Rockaway Park Shuttle.

Track connections from the yard connect both railroad north to Euclid Avenue and railroad south past Grant Avenue on the IND Fulton Street Line. This allows trains to be added or removed from service in either direction. 40°40′21″N 73°51′54″W

Grant Avenue entrance of Pitkin Yard

Rockaway Park Yard

Rockaway Park Yard is located in Rockaway Park, Queens. It is a layup yard for the R46s on the Rockaway Park Shuttle and **A** train, although they are primarily maintained at the Pitkin Yard in Brooklyn. R40A #4392-93 are used for school training in this yard.

Rockaway Park Yard eastern part

Western end of Rockaway Park Yard

This yard is adjacent to Rockaway Park – Beach 116th Street. Like the IND Rockaway Line itself, the Rockaway Yard was originally a yard for the Rockaway Beach Branch of the Long Island Rail Road. It included a water tower, a roundhouse, and an elevated loop track for Brooklyn Rapid Transit trains. 40°34′52″N 73°50′8″W

Other Yards

36th-38th Street Yard

The **36th–38th Street Yard** is located between Fifth and Seventh Avenues in Sunset Park, Brooklyn, adjacent to the Jackie Gleason Bus Depot. This yard is not normally used for revenue-service train maintenance, though some trains for the **R** service are stored here. Its primary function is to store diesel and electrically powered maintenance-of-way and other non-revenue service rolling stock. It is also used to transfer trash from garbage collector trains to trucks via platforms inside the yard just south of 37th Street.

North side of 36th-38th Street Yard

This southern part of the yard was formerly the center of the South Brooklyn Railway, which extended from Bush Terminal through the north part of the yard, then down Gravesend Avenue and into the Coney Island Yard. The yard is entirely equipped with hand-operated switches. Only Fresh Pond Yard and East New York Yard share this characteristic. 40°38′52″N 73°59′48″W

Linden Shops

Linden Shops is a track shop, where track switches and other components are assembled. It has track connections to the IRT New Lots Line and BMT Canarsie Line, but no third rail, restricting the facility to diesel-powered trains only. There is also a track connection to the Long Island Rail Road's Bay Ridge Branch. This connection is one of two from the subway to the mainline United States rail network (the BMT West End Line is the other). 40°39′13″N 73°54′16″W

References

[1] "East 180th St. & Unionport Yards" (http://www.nycsubway.org/cars/yards_180th.html). . Retrieved 2012-03-01.

[2] IRT Subway: Small action at the (2) (5) Unionport Yard (180th Street) http://www.youtube.com/watch?v=i5_KNc0Mk5Q&

[3] "The Bronx IRT, Lenox/White Plains Road/Dyre Avenue Line" (http://www.nycsubway.org/lines/whiteplains.html). . Retrieved 2012-03-04.

[4] "National Register Information System" (http://nrhp.focus.nps.gov/natreg/docs/All_Data.html). *National Register of Historic Places*. National Park Service. 2009-03-13. .

[5] Kathleen A. Howe (March 2004). "National Register of Historic Places Registration: Concourse Yard Entry Buildings" (http://www.oprhp.state.ny.us/hpimaging/hp_view.asp?GroupView=101482). New York State Office of Parks, Recreation and Historic Preservation. . Retrieved 2011-01-12. *See also:* "Accompanying five photos" (http://www.oprhp.state.ny.us/hpimaging/hp_view.asp?GroupView=101483). .

[6] "East New York Yard" (http://www.nycsubway.org/cars/yards_eastny.html). NYCSubway.org. . Retrieved 2012-03-02.

[7] Track map created by Peter Dougherty (http://images.nycsubway.org/trackmap/detail-71-cont.png) and published in his book "Tracks of the New York City Subway" (Fourth Edition)
[8] Richmond Hill Record (http://www.richmondhillhistory.org/BattleofLI.html)
[9] UTU Article 2004 (http://www.utu.org/worksite/detail_news.cfm?ArticleID=17348)
[10] Kennedy, Randy (October 27, 2004). "A Day in the Subway, as It Rolls Up a Century" (http://www.nytimes.com/2004/10/27/nyregion/27subway.html). *New York Times*. . Retrieved September 2, 2009.
[11] MTA, Capital Program 2005–2009 (http://www.mta.info/mta/capital/index.html), accessed April 17, 2007

External links

- Subway Yards (http://nycsubway.org/cars/yards.html)

Article Sources and Contributors

T_(New_York_City_Subway_service) *Source*: http://en.wikipedia.org/w/index.php?title=T_%28New_York_City_Subway_service%29 *Contributors*: 2001:DA8:D800:701:60BA:D83B:37B2:AC74, Acps110, Alphachimp, Amalas, Beagel, Bennyp77, CComMack, Cecropia, ChrisRuvolo, Cnbrb, Dcandeto, Eco84, Eric B, Flamurai, Gfoley4, Gimmetrow, Hgrosser, Hinto, IGeMiNix, IRT1904, Jc42bugs, Jeffknj8, Jimmy Slade, KKane9, Ken S., Larry V, Marc Shepherd, NE2, Newkirk Plaza David, Omicronpersei8, Pacific Coast Highway, R68A B Train, Rapomon, SPUI, SchuminWeb, The Legendary Ranger, Tinlinkin, Trulystand700, Vcohen, Vrysxy, 75 anonymous edits

New_York_City_Subway *Source*: http://en.wikipedia.org/w/index.php?title=New_York_City_Subway *Contributors*: 125t, 1995hoo, 212tumble, 2600:1011:B02F:A477:DDC9:6F57:AC17:BE53, 7Train, A3RO, ABehrens, AEMoreira042281, AMLNet49, ATren, Abba701, AbigayilNeshama, AbsolutDan, Acps110, Adam mugliston, Adolphus79, Akkida, Al E., Alai, Alansohn, Ale cyn, AlexRampaul, Alexissheridan, AlistairMcMillan, Allamericanbear, Alphachimp, Alsandro, AmontonarPapeles, Andrew Levine, Andrew from NC, Andrewpmk, Andy Marchbanks, AndyVolykhov, Andyinuk17, AnonMoos, Anonblint2, Apricot50, Argon233, Armadillo1985, Armyatoms10, Arsenikk, Artfriend2008, Ashanda, Astuishin, Aude, Auric, Aurora30, Autopilot, Avman89, Avocado, Aydiosmio, Azniiinvazn, BD2412, BOARshevik, BWCNY, Babyox4420, BalticPat22, Beetstra, Before My Ken, Benandorsqueaks, Bender235, Bentley4, BernardM, Bernstein2291, Bhludzin, Bigrene2, Bill william compton, Birdhurst, Bkell, BlastOButter42, Bleakcomb, Blue387, Bluerattle, Bmclaughlin9, Bob schwartz, Bolwerk, Bonadea, Boodleheimer, Borromakot, BostonMA, BrOnXbOmBr21, Brandonh, BrianAdler, Burch003, Burgundavia, C.Logan, CComMack, Cacophony, Calabe1992, Calton, Canadaolympic989, Cantara, Canterberry, Capricorn42, Cassowary, Castncoot, Cecropia, Chaohwa, Charles01, CharlesHBennett, Che829, Chowbok, Chuck Sirloin, Ckatz, Cleduc, Clevelandguy, Clicq, Clubjuggle, Clukic, Cmdrjameson, Cocoaguy, Coemgenus, Colonies Chris, Commandr Cody, CommonsDelinker, Comwhiz2002, Concoursegrl, ContinentalAve, CoolGuy, Corti, CountZ, Cremepuff222, Crxmen, Cyrius, DMG413, DMacks, DReifGalaxyM31, DVdm, Dale Arnett, Damian Yerrick, DanTD, DanTheMan702, Darkwind, DarwinPeacock, Daveny1979, David Eerdmans, David Shankbone, DavidSpanel, Davija, Dcandeto, Dcox10, Deor, DerHexer, Deror avi, Descent, Desernio, Deuce1980, Devoxo, Dhartung, Dhnlin, Dicklyon, Dillan.Murray, Dispenser, DjHammers, Donald Albury, Donald Friedman, Donherbert, Doseiai2, Drago9034, Dream out loud, DrumIntellect, Dschuster17, Dschwen, DubbaG, Dukeofomnium, Dumbpigeon, Dynzmoar, Dysprosia, Eccles71B, Ekozie, ElBenevolente, Elmer Clark, Enfieldfalls, Enviroboy, Epbr123, Epeefleche, Epicgenius, Eraserhead1, Ericamick, Error46146, Esrever, Evgenior, Ewlyahoocom, Extrahence100, Fairlyoddparents1234, Fan Railer, Farmer77, Farras Octara, Fconaway, Flamurai, Foobody, FourthAve, Frankie816, Friedo, Frietjes, Fuhghettaboutit, Gail, Gaius Cornelius, GeeJo, Gemini1980, Geoking66, Georgejoyrider, Gfoley4, Ghostbusters in NY, Gjs238, Glen, Gnusbiz, Gordeonbleu, GraemeL, Grahamec, Grammarcop1, GregorB, Grenavitar, Gridlock Joe, Grimitar, Ground Zero, Grouse, Guoguo914, Gurch, Gwernol, HManat32, Hairymon, Hajor, Hbdragon88, HereToHelp, Herenthere, Herostratus, Hertz1888, Hex, Hgrosser, Hinto, Hmains, Hockeyman001, Hux, I.M. Rich, IGeMiNix, II MusLiM HyBRiD II, IIVQ, IND Second System, IRT1904, Ich weiß es nicht, Imdanumber1, Imroy, Insanity Incarnate, Insuranze, Inter16, Iridescent, Izehar, J 1982, J.delanoy, J4lambert, JNW, JONJONAUG, Ja 62, Jackehammond, Jaedglass, Jalabi99, Jason M, Jaxl, Jcembree, Jenny Wong, Jesse Viviano, Jessekeim, Jfruh, Jim856796, Jimmy Slade, Jklamo, Jleon, Jmcdon10, Jmchuff, John, John of Reading, Johntex, Josephfg, JoshuaZ, KOknockout920, Kaaveh Ahangar, Kbdank71, Kelvinc, Keon Milton, KerathFreeman, Ketiltrout, Kgwo1972, Khazar2, Kildor, Kinkyturnip, Kirchsw, Kitsonk, KnightSabre 3, Koavf, Kokin, Korny O'Near, Krallja, Krashlandon, Kribbeh, Kthapelo, Kurumban, LLarson, La goutte de pluie, Lampbane, Larry V, LarryMorseDCOhio, Lengli, Leslie Mateus, Level Crossing, Levg79, Levineps, LibrarianB, Lightmouse, Likelife, LilHelpa, Llamabr, Locuteh, Loodog, Lord Voldemort, LoreMariano, Lost on belmont, Lotje, LouI, Luminifer, M.O.X, MJCdetroit, MJofLakeland, MOOOOOPS, Macrakis, Madhero88, Mahjongg, Mailer diablo, Makeemlighter, Marc Shepherd, Massysett, Matthew Yeager, Mattweinberg, Mauls, Maxis ftw, Maxo, Mazeau, Mbssbs, MegX, Melanarchy, Messy Thinking, Metamorphosis, Metstotop333, Michael Hardy, Miketdt, MileMarker651, Millbrooky, Mindspillage, Modest Genius, Modster, Modulatum, Mogism, Mohammedjanich, Moreau36, MrFish, MrLincoln, Mschamis, Mtattrain, Mulad, Multivariable, Myhlow, NE2, Natalie Erin, Natl1, Neilc, Nelson50, Neutrality, Newkai, Newkirk Plaza David, Newyorkbrad, Nfitz, Nick22, Nickj, Niffweed17, Night Gyr, Nightscream, Nightstallion, Niteowlneils, Norm mit, Nricardo, Nuggetboy, Nycalex2012, Nyctransitgov, Nytix, O.Koslowski, Oanabay04, Oatmeal batman, Occur Curve, Oconli1098, Ohconfucius, Ohnoitsjamie, Oknazevad, Onionmon, Optimusnauta, P. S. Burton, PZFUN, Pacific Coast Highway, Pah246, Paktype, Palica, Patcat88, PaulGS, PeeJay2K3, Pegship, PennaBoy, Penser, Peter Delmonte, Peter Horn, PeterEarlJorgensen, Pgecaj, Pharos, Phoenix-wiki, Pi.1415926535, PimRijkee, Pinsch, Plasma east, Plasticup, Pne, PoccilScript, Pokey54, Pomte, Poorpete, Postdlf, Potsy233, Producercunningham, Professoracc, Président, PseudoSudo, Publicroutes, Quarl, Queensdude, Quentin mcalmott, R36, REDJASPER777, RRKennison, Rackham, Radio Guy, Radiojon, Radon210, Rahaeli, Ranever, Raryel, Reach Out to the Truth, Reck345, Red Card For You, Red Winged Duck, RedBLACKandBURN, Reddi, Redmagic55, Redthoreau, Reluctantpopstar, Remember the dot, Renek, RevelationDirect, Reywas92, Rich Farmbrough, Richiekim, Rickyrab, Ringbark, Rjwilmsi, Rlendog, Rnpedersen18, Roadcrusher3, Roadrunner, Roadrunner3000, Rockies77, Rogerd, Ronhjones, Rsd315, SDC, SNIyer12, SPUI, STE-HTM, Sammetsfan, Sanchito, Sarahstern, Sbrgc, Schaengel89, SchreiberBike, SchuminWeb, Schweiwikist, ScottyBerg, Scrabbleship, Screamingman14, Sdalmonte, Sdedeo, Seeadam, Seidenstud, Ser Amantio di Nicolao, Sg09836, Sgw1009, Shadowjams, Shakazulu8999, Shlomke, Shuki, Sift&Winnow, SingCal, Skarebo, Skew-t, Skybon, Slambo, SliceNYC, Sligocki, Smartyllama, Soarhead77, Someaznguyny, Sonicyouth86, Soupnyc807, SpiceMan, Sport and politics, Spudnyc, Squigish, Srittau, Sshikora, Steevo714, Stevenghetto257, Stoogeyp, Storm2k, Student7, Stwalkerster, Sunstar, Svgalbertian, Sweek, Swessdog, Symane, TMV943, TPIRFanSteve, Taco, Tagishsimon, Tanweer Morshed, Tarquin, Tasc, Taylorluker, Tc61380, Telekid, Tennisuser123, Tevi, Tex, Tgeairn, Tgerr55, Thandelike, The Anome, The Legendary Ranger, The Thing That Should Not Be, TheSuave, Thefirstchoice, Thingg, This51man, ThisIsMyWikipediaName, Thistheman, Tholme, Thos Davis, Thumperward, Tide rolls, Timverma, Tinlinkin, Tlogmer, Tobyvonmeistersinger, Tom X. Tobin, TomTheHand, Tomatosoop99, Tommy2010, Tomokeefe, Tony Sidaway, Tony1, Toytoy, Train2104, Transit2000, Trulystand700, Truthanado, TypoDotOrg, URRamblings, Ufwuct, UltimatePyro, Untogethered, UpstateNYer, Uris, VMIKEW, Vancouverguy, Vanka5, Vcohen, Vegaswikian, VerruckteDan, ViaJFK, Vicki Rosenzweig, Vince718, Vmenkov, VolatileChemical, Wanderingstan, WarriorGoddess, Wavelength, Wereon, WhisperToMe, Wikiacc, Wikidudeman, Willking1979, Wmahan, Wongm, Woohookitty, Wv235, Yellowdesk, YibiaoWang, Ylee, YodaRULZ, Yorker, You are encouraged to log in, Yukon depot, Yulu, Yussi1870, Zack, Zanimum, Zawar1, ZekeMacNeil, Zmartine, Zoicon5, Zondor, ZooFari, Zzyzx11, 749 anonymous edits

BMT_West_End_Line *Source*: http://en.wikipedia.org/w/index.php?title=BMT_West_End_Line *Contributors*: Acps110, Axpde, Cecropia, DanTD, EddieSegoura, Elipongo, Espresso Addict, Ethan2226, GK tramrunner, Jim.henderson, Keon Milton, Larry V, Levg79, Marc Shepherd, Marc omorain, Modster, NE2, OP Punster, Platzmania, Playclever, SPUI, Skysmith, Stevenghetto257, The Legendary Ranger, Tinlinkin, Train2104, Vcohen, Vrysxy, 16 anonymous edits

Second_Avenue_Subway *Source*: http://en.wikipedia.org/w/index.php?title=Second_Avenue_Subway *Contributors*: AJC340, Acps110, Agiorgio, Alansohn, Alsandro, Amakuru, Anetode, AnonMoos, ArnoldReinhold, AtomikWeasel, Atrix20, BWCNY, Badger151, Barnt001, Bdentzer, Bdiscoe, Beagel, Before My Ken, Bender235, Blauwkoe, Bmitchelf, Brianga, Bry9000, Bvitamini, CComMack, CanisRufus, Cbastian, Cecropia, ChrisRuvolo, CliffC, Cmprince, Cnbrb, Colonies Chris, CoolGuy, Cxarli, D Monack, DMG413, DMacks, DanTD, DarwinPeacock, DavidRF, Daviessimo, Dcandeto, Digitat, Dlauri, DocWatson42, Dream out loud, Dryamaka, ELMtree, Edg2103, Epeefleche, Ethelh, Eubene, Ewlloyd, FFM784, Flamurai, FourthAve, Friedo, GGreeneVa, Gaius Cornelius, Gatewaycat, Gc11530, Georgejoyrider, Gfoley4, Giraffedata, Grenavitar, Gwernol, HManat23, HManat32, Harddavid, Hbdragon88, Hemlock Martinis, Herenthere, Hmanck, Howcheng, Ibagli, JamesAM, Jayjg, Jd2718, JeremyA, JesseRafe, Jim.henderson, Jim1138, Jimmy Slade, Jivecat, Jkeene, Jmcdon10, Jmchuff, John Silvestri, JonathanDP81, JulieADriver, JustSomePics, Kakoui, Kbdank71, Keepweek90, Kelvinc, Kildor, Kjordahl, Kkmurray, Kolega2357, Lanoitarus, Larry V, Lensovet, Lentower, Magog the Ogre, Marc Shepherd, Massysett, Matthiashess, Me Three, Mfk91, Modster, Moncrief, Mr. Laser Beam, MrLincoln, Mrcatzilla, Mtattrain, Murrows, Muttonking, Mynameinc, NE2, NTK, NYCRuss, Nbarth, NcSchu, Ng.j, Niffweed17, Nightstallion, NobleHelium, NuttyJay, Nv8200p, Oanabay04, Oknazevad, Oneworld25, PROSA, PTSE, Pacific Coast Highway, PaulinSaudi, Peace01234, Peter Horn, Phil Bastian, Philg88, Plaws, Postdlf, Psantora, R'n'B, Radio Guy, Raryel, Reluctantpopstar, Revas, Rhg, Rich Farmbrough, Rickyrab, Rjanag, Rjwilmsi, Roadcrusher3, Roadcrusher4, Robert Weemeyer, Robert1947, Rock nj, Rockies77, RoySmith, Rwalker, RyanGerbil10, SPUI, Sam Blacketer, Sbmehta, SchuminWeb, ScottyBerg, Screamingman14, Slambo, Smackdown10, Ssilvers, Stevenghetto257, Sunaryo Joyopuspito, Swarve, Sylvain1972, TenOfAllTrades, The Legendary Ranger, Thehornet, Timc, Tinlinkin, Titanosaurus, Tlogmer, Tomatosoop99, Train2104, Transpoman, Trink24, Trulystand700, Truthanado, UFu, Ua747sp, Utcursch, VT hawkeye, Vcohen, Vladmirfish, Vrysxy, WCCasey, Wikih101, Woodshed, Wv235, Y, Yellowdesk, Zenyu, Zoicon5, 192 anonymous edits

The_Bronx *Source*: http://en.wikipedia.org/w/index.php?title=The_Bronx *Contributors*: 00jdaley, 0o-JayParmar-o0, 1122334455, 195.186.144.xxx, 1Sire, 22dragon22burn, 28421u2232nfenfcenc, 5 albert square, 66.65.57.xxx, 83748a017385985739, 9258fahsflkh917fas, A. Carty, AEMoreira042281, Aarchiba, AardvarkOfHate, Aaronbrick, Acalamari, Accurizer, Acntx, Acps110, Addshore, Aff123a, Agold2121, Ahoerstemeier, Aitias, AjaxHegarty, Al Silonov, AlanUS, Alaniaris, Alansohn, Alex '05, Alex Usoltsev, AlexiusHoratius, AlistairMcMillan, Almartin82, Alsandro, Alucard (Dr.), Amc382, Amorrow, Anders Jönsson, Andopento, Andre Engels, Andy Marchbanks, Andycjp, Angelrendon, Angr, Ankyese Csuk, Antandrus, Anthony22, Anthonyramos1, Antipastarasta, Antony-22, Anubis2051, Apollo1758, Aralvarez, Arden, Arse08, Art LaPella, Artrix, Ashley Pomeroy, Astuishin, Atlant, AuburnPilot, Aude, Autosol, BD2412, BRG, Badger Drink, BaronLarf, Basser g, Bdcrph, Bearian, Beetstra, Before My Ken, Beige Tangerine, Beland, Bellagio99, Bellatrix Kerrigan, Bender235, Benjwong, Bentley4, BernardM, Besselfunctions, Bfigura's puppy, Bigchrisk919, Biglu31, Bill Davis, Bill37212, Bjones, Bjsamelsonjones, Blablabla129, BlackNYer, Bletch, Blethering Scot, Bloommolly, Bmf 51, Bo Lindbergh, Bobblewik, Bolwerk, Bonadea, Bongwarrior, Bradv, BrendelSignature, Bronks, Bronxcounty, Bronxcountygov, Bronxrap, Brooklyn NYC, Brossow, Bssc81, Bubba hotep, Buckner 1986, Buddha24, Burnwelk, Bxsstudent, CComMack, CHJL, CORNELIUSSEON, CSZero, CUSENZA Mario, Cadiomals, Cafzal, Calineed, Camelbinky, Can't sleep, clown will eat me, Canadian-Bacon, CanadianLinuxUser, Canis Lupus, Carlos10457, Carlosguitar, CaseyPenk, Catbar, CatherineMunro, Cdang, Cdc, Cfortunato, Chairman S., Chcknwnm, Checkmatechamp137, Chicago103, ChildofMidnight, Chillywillycd, Chocolatechipcookie91, Chowbok, Chris Buckey, Chris the speller, ChrisRuvolo, ChristTrekker, Chuiknows2, Citidm, Ck lostsword, CliffC, ClockworkSoul, Cmr08, Cobaltbluetony, Cocytus, Colfer2, Colonies Chris, Cometstyles, CommonsDelinker, Conny, Contributor777, Conversion script, Coolguy82, Coreybower, Crazynas, CubBC, Curlymeatball38, Cwkmail, Cxawer, Cybercobra, Czaruno, DBaba, Dainamo, DanKeshet, DangApricot, Danielgrenell, Darth Naeblis, David Kernow, David Shankbone, DavidLevinson, DavidSpanel, Dbo789, Dcs315, Deathardath, Debresser, Deckoffa, Decltype, Decumanus, Deltopia, Deor, Dermann69, Deus257, Devtrash, Dfmock, DigiBullet, Dissent, Djflem, Dmwime, Doc9871, DocWatson42, Doceddi, Dogears, Dogru144, Dolovis, Dragoon009, Drawn Some, Dreamyshade, Druzhnik, Dtasripin, Duffman, Dycedarg, DyluckTRocket, Dysprosia, EWS23, Earlpwns, EastCoastGhost, Eastlaw, Echuck215, Eco84, Ed g2s, Edgar181, Eduardo Sellan III, EdwinCasadoBaez, Egil, EhJJ, Ehlkej, Elipongo, Elmoelmo, Emeraldcityserendipity, Emeritahears, Emigrant85, Epbr123, Erianna, Eritasactium, Erose915, Euphen, Evans1982, Evercat, Excirial, Explicit, Exxy, FSG, Fabhcún, Faithlessthewonderboy, Falconus, Farras Octara, Ferengi, Ferkelparade, FinanceANDinvesting, Fixifex, Foobarnix, Forsse90, Fortdj33, Fraggle81, Fraterm, Frecklefoot, Free Bear, Frog47, Froid, Frozen4322, Fuhghettaboutit, Funnyhat, Furries, Futurebird, Gadfium, Gaius Cornelius, Gareth E Kegg, Gargolla, Gary King, Gayest, GeorgeLouis, Gerrish, Gerry D, Gettingtoit, Gidonb, Giuan, Gjs238, Glover, Golbez, Good Olfactory,

GordyJ, GotNoGutsGotNoGlory, Graham87, Greg Tyler, GregorB, Ground Zero, Groupuscule, Gtrmp, H1nkles, HalfShadow, HannahMiley, Hari akku, Harmsma, Headbomb, Hemmingsen, Herr Ratselhaft, Higgs11927, Hitmanmj3, Hmcnally, Hochung4, Hoverfish, IRP, Iago Dali, Ian Fortuno, Iancurtis86, Ibukazi, Icarusgeek, Icenine378, Imran, Infernoacid, Infrogmation, Isfisk, It Is Me Here, Italiano soprano, Itisbig, Iwinick, Ixfd64, J.delanoy, JBC3, JBsupreme, JNW, Jackstraw00, Jacob Myers, James100, Jameskworth, Jamesontai, Japanese Searobin, Japo, Jasonglchu, Jaysscholar, Jclemens, Jd2718, Jeff G., Jeffhoy, Jerk in a hat, Jersey Devil, Jerseyj00, Jfalk, Jhsounds, Jiang, Jim.henderson, JimWae, Jimsteele9999, JinJian, Jleon, Jlevy, Jlschaefer, JoeBrennan, Joelbrandw, Joelwiki33, John of Reading, John254, Johnbrownsbody, Johnedwineckert, Johnpacklambert, Jomasecu, JonC0001, JonathanDP81, Jonik, Jonnyfishboy, Joowwww, JordoCo, Jorunn, Josemanimala, Joseph Solis in Australia, Josephlol11, Jrdodge, Juicybuoy, JustAGal, K72ndst, KGasso, KOknockout920, KXL, Kaaveh Ahangar, Kaisershatner, Kajaro, Kalamkaar, Karenjc, Kbir1, Keilana, Keon Milton, Kevin Rector, KingBee1974, Kingturtle, Kitty53, Kjaergaard, Kmccoy, Knepflerle, KnowledgeOfSelf, Kollision, KoopaKid17, Krallja, Krockblt515, Krrmw, Ktr101, Kubigula, Kuru, Kusunose, Kyle1278, LUUSAP, Lampbane, Larry V, Lauracs, LeRoi, LedgendGamer, Leeheaven, Lego872, Leo44, Lester Long, Levineps, LewisMorris, Liammarkrickard, Life of Riley, Lightmouse, Lilac Soul, Lolwutnub, Lord Emsworth, Lord Voldemort, LostAccount, LuK3, MBernal615, MJBear, MJBurrage, MJCdetroit, Mad6986, Magioladitis, Malcolm bg, ManPigBread, Mandarax, Manop, Mapple, Markustwofour, Martian.knight, Mattbr, Mav, Mayumashu, Mazeartist, Mbg6231, Mboverload, McGrupp10799, Mcbgirl, Mdekadt, Meaghan, Medeis, Mediapr, Medicalfools, Mentifisto, Metropolitan99, Mfk91, Mghabmw, Michael L. Kaufman, MichaelBillington, MichaelMcCain, Michelle eris, MidnightCrisis, Mike Rosoft, Mike6271, Milton Stanley, Minecraft1997, Minimac, MisterPitt, Mlaffs, Mm40, Modernist, Modster, Moisture, Moncrief, Moosicus, Morgan Cromwell, Mr Rhys, MrB, MrVibrating, Mrs peg, Ms2ger, Mschlindwein, Muboshgu, Mwanner, Mynameinc, NE2, NHRHS2010, NJ hardcore kid, NYCRuss, NYCfellow, Narture, NawlinWiki, Neil916, Nepenthes, Netnuevo, Neutrality, Newkirk Plaza David, Nick Levinson, Nimish Gautam, Nine11c2, Nivix, Nkabouris, Nlsanand, Noq, Nothingofwater, Nricardo, NuclearWarfare, Nuggetboy, Numbo3, Nyttend, Ohnoitsjamie, Oknazevad, OlEnglish, Oleg Alexandrov, Olego, Ottawa4ever, P.D., P0per, PGWG, PKT, Pacific Coast Highway, Paganimage, Patrick, Patrickthompson07, PaulHanson, PaulLev, Paulmlieberman, Paultrimm, Pb30, Pbogomiakov, Per Ardua, PeterJohnson, Petri Krohn, Pgecaj, Pharos, Phil Boswell, Philip Trueman, Phillyboy2204, Pigman, Pinethicket, Pinklink98, Piperdown, Pookman7497, Poopc, Possum, Postdlf, Power level (Dragon Ball), PranksterTurtle, President David Palmer, Professor water, Project runway06, Pseudomonas, Pstanton, Pubdog, PumpkinSky, Pvita, Quadell, Quantumor, Qxz, R'n'B, R.T.Gellar, R9tgokunks, R@y88, RainbowOfLight, Rally 21, Ram-Man, Rama, Random Passer-by, RandomP, Rayshi1, Reconfirmer, Red Deadeye, Regan123, RegentsPark, Retired username, RetiredWikipedian789, Revas, Rex Germanus, RexNL, Rhobite, Rich Farmbrough, Richiekim, Rick Block, Rickterp, Rjensen, Rjwilmsi, Rmm289, Robbie098, Robo Cop, RockJuno, Rocketrye12, Rockstable, Rod121991, Rogerd, Rooflee, RoseParks, RossPatterson, Rownak, RoySmith, Rrburke, Rtd2101, RunOrDie, Ryanaxp, Ryomaandres, SDC, SGT141, SJP, SNIyer12, ST47, SWick1973, Sandy June, Satori Son, Satyadasa, SaxofoneDL, Scanlan, SchnitzelMannGreek, ScottyBerg, Seaphoto, Seduisant, Seidenstud, SelfStudyBuddy, Senty78, Sexyasssonja19, Sfan00 IMG, Sha83, Shakescene, Sherool, Shoreranger, Shunland, Simeon H, Simpatico, Simslord06, Sisterdetestai, Sk5893, SkyWalker, Skyline1724, Slarre, Smallbones, Smooth O, Sonicology, South Bronx 2000, Sparkicus, Special-T, Srbruner, Sswonk, Station1, Steeldx, SteveCrook, SteveSims, Stevenrasnick, Stormie, SummerPhD, Supercool Dude, Surge79uwf, Swarve, TOttenville8, TPIRFanSteve, Tantris, Taroaldo, Tarret, Tascha96, Tassedethe, Tatianahinterreither, Taz82189, Tbhotch, Tbone874, Tchronopoulos, Template namespace initialisation script, Tenebrae, Tex, The Anome, The Emirr, The Interloafer, The Universe Is Cool, TheBronxNYC, Theaterfreak64, Theguv2000, Thejaggedtimes, Thejimbaker, Themfromspace, Theopolisme, Therecanonlybetwo, Thingg, Thumperward, Tiburon, Tifoo, Tim Sailor, Timneu22, Tintero, Titanosaurus, Tkynerd, Tlogmer, Tobby72, Toesboy13, Tomwhite56, Topbanana, TreasuryTag, Treezum, Treisijs, Tvfan01nine, Tvoz, Twsx, Ucucha, Ufwuct, Ukabia, Ulric1313, Universearle, Usability 5, UtherSRG, VanCortlander1, Vanished user 5zariu3jisj0j4irj, Varlaam, Vary, Vegaswikian, Vicki Rosenzweig, Vilerage, Voidvector, W guice, Waninge, Wars, Wasted Time R, Wavelength, Wbennin, Wegesrand, Weitzlux, Wetman, WhisperToMe, WikHead, Wiki Roxor, Wikiwiki718, WillC, Willywonka3505, Wimt, Wintran, Wizard1022, Wizard98, Wknight94, Wolfgang Hagen, Woodieburg, Woodlawn1977, Woohookitty, Wv235, X!, XJIMMYBOYX, XLR8TION, Xerius, Xiaodown, Xompanthy, Xpendersx, Xqt, Y2kcrazyjoker4, Yahel Guhan, Yamamoto Ichiro, Yekrats, Yesopenilno, Yidisheryid, Ytny, Yurivict, Zaf, Zaljaura2, Zaynmalik3331, Zcastro25, Zeamays, Zidonuke, Zoicon5, Zumwalte, Zweifel, Zyxw, ^demon, , 1253 anonymous edits

125th_Street_(Manhattan) *Source*: http://en.wikipedia.org/w/index.php?title=125th_Street_%28Manhattan%29 *Contributors*: Acps110, Alansohn, Americasroof, AndrewHowse, Bachrach44, Beland, Candleabracadabra, Censorwolf, Cjs2111, Clarkcj12, Dananderson, Deror avi, Efb91, Gavia immer, Grutness, IP 66.173.44.202, Jak123, Jfioeawfjdls453, Jim.henderson, Jjc104, Joke137, Jpbowen, KlooBawt, Knee427, Larry V, Lfh, Mairi, NYCfellow, NameIsRon, Pacific Coast Highway, PaulHanson, Petri Krohn, Pharos, RandomXYZb, Rjwilmsi, SPUI, Schlaier, Shakrat, Siroxo, Srikipedia, Station1, The Legendary Ranger, Tony1, Transfinite, Transpoman, TwinsMetsFan, Vgranucci, Vranak, WhisperToMe, Ytny, 29 anonymous edits

List_of_New_York_City_Subway_lines *Source*: http://en.wikipedia.org/w/index.php?title=List_of_New_York_City_Subway_lines *Contributors*: Acps110, AndyVolykhov, CComMack, Cecropia, Dhnlin, Frietjes, I.M. Rich, J intela, Jd2718, Larry V, Marc Shepherd, Mirom6, NE2, Niffweed17, Nightstallion, Oknazevad, RyanGerbil10, SPUI, Tinlinkin, Vcohen, Vrysxy, Woohookitty, 12 anonymous edits

Unused_New_York_City_Subway_service_labels *Source*: http://en.wikipedia.org/w/index.php?title=Unused_New_York_City_Subway_service_labels *Contributors*: AMLNet49, Acps110, Badger151, C.Logan, Cecropia, CrazyC83, DCarltonsm@msn.com, DMacks, DReifGalaxyM31, Dhnlin, Dough4872, Dream out loud, Dudi144, Eric B, Fennec, IGeMiNix, IRT1904, Jimmy Slade, Jkeene, Larry V, Marc Shepherd, Mild Bill Hiccup, Mindmatrix, Mostly Rainy, MrLincoln, Mtattrain, NE2, Newyorkbrad, Oknazevad, Phlogistomania, R36, Roadrunner3000, SPUI, SchuminWeb, Skipstops, The Legendary Ranger, Thecornkid, Tinlinkin, Train2104, Transpoman, Trulystand700, Vcohen, Vrysxy, Will Beback Auto, Yun-Yuuzhan (lost password), 75 anonymous edits

Second_Avenue_(Manhattan) *Source*: http://en.wikipedia.org/w/index.php?title=Second_Avenue_%28Manhattan%29 *Contributors*: 16@r, Acps110, Alansohn, Artegorov201, BD2412, Beyond My Ken, C16sh, David Shankbone, Deror avi, Dlindner, Funandtrvl, George Burgess, Godfreychristian, Grutness, Gsd97jks, Hmains, Iloveparis, IronGargoyle, Jeff G., Jim.henderson, Kafziel, LHS, Leifern, Lovelac7, MaGioZal, Moncrief, NE2, Nareek, Oshea41, Pradiptaray, Robertknyc, SPUI, Schlaier, ScottyBerg, Station1, T-bomb, TIY, The Canadian Roadgeek, Transpoman, Vis-a-visconti, Vrysxy, , 12 anonymous edits

List_of_New_York_City_Subway_inter-division_connections *Source*: http://en.wikipedia.org/w/index.php?title=List_of_New_York_City_Subway_inter-division_connections *Contributors*: Acps110, Alai, Marc Shepherd, NE2, Nightstallion, RyanGerbil10, SPUI, Vcohen, Vrysxy, 10 anonymous edits

New_York_City_Subway_yards *Source*: http://en.wikipedia.org/w/index.php?title=New_York_City_Subway_yards *Contributors*: Acps110, Bebo2good1, DanTD, Ethan2226, IGeMiNix, Imdanumber1, Jim.henderson, Joeal532, Larry V, Marc Shepherd, NE2, Oknazevad, SMasters, SPUI, Salvidrim, SkipperRipper, The Legendary Ranger, Train2104, Tvega961, Vcohen, Vladmirfish, Vrysxy, Wusel007, Zimbabweed, 74 anonymous edits

Image Sources, Licenses and Contributors

file:NYCS-bull-trans-T.svg *Source*: http://en.wikipedia.org/w/index.php?title=File:NYCS-bull-trans-T.svg *License*: unknown *Contributors*: User:Flamurai

File:2nd ave subway.svg *Source*: http://en.wikipedia.org/w/index.php?title=File:2nd_ave_subway.svg *License*: unknown *Contributors*: 21:12, 9 July 2008 (UTC)

File:MTA New York City Subway logo.svg *Source*: http://en.wikipedia.org/w/index.php?title=File:MTA_New_York_City_Subway_logo.svg *License*: unknown *Contributors*: Metropolitan Transportation Authority

File:NYC Subway R142 on the 4 R160A on the E.jpg *Source*: http://en.wikipedia.org/w/index.php?title=File:NYC_Subway_R142_on_the_4_R160A_on_the_E.jpg *License*: unknown *Contributors*: Level Crossing

File:NYC Subway Times Square.jpg *Source*: http://en.wikipedia.org/w/index.php?title=File:NYC_Subway_Times_Square.jpg *License*: unknown *Contributors*: Beyond My Ken, Flamurai, Gryffindor, Infrogmation, NE2, Quasipalm, SPUI, Uris, Wolfmann, Ytoyoda

File:Interborough Rattled Transit Restored.png *Source*: http://en.wikipedia.org/w/index.php?title=File:Interborough_Rattled_Transit_Restored.png *License*: unknown *Contributors*: Rogers, W. A. (William Allen), 1854-1931, artist.

Image:Nuvola apps kaboodle.svg *Source*: http://en.wikipedia.org/w/index.php?title=File:Nuvola_apps_kaboodle.svg *License*: unknown *Contributors*: Pierpao, Tkgd2007, Waldir, 1 anonymous edits

File:1906 IRTC map South.png *Source*: http://en.wikipedia.org/w/index.php?title=File:1906_IRTC_map_South.png *License*: unknown *Contributors*: DarkEvil

File:Cortlandt St station demolished.jpg *Source*: http://en.wikipedia.org/w/index.php?title=File:Cortlandt_St_station_demolished.jpg *License*: unknown *Contributors*: Acps110, DanTD, Dream out loud, Train2104

File:New York City Subway.png *Source*: http://en.wikipedia.org/w/index.php?title=File:New_York_City_Subway.png *License*: unknown *Contributors*: Gary Mark Smith

File:Subway elevated2.jpg *Source*: http://en.wikipedia.org/w/index.php?title=File:Subway_elevated2.jpg *License*: unknown *Contributors*: User:Dschwen

File:R142 Sign.jpg *Source*: http://en.wikipedia.org/w/index.php?title=File:R142_Sign.jpg *License*: unknown *Contributors*: Johnny S., Pacific Coast Highway, Vcohen, Ytoyoda

File:NYCS-bull-trans-1.svg *Source*: http://en.wikipedia.org/w/index.php?title=File:NYCS-bull-trans-1.svg *License*: unknown *Contributors*: User:Flamurai

File:NYCS-bull-trans-2.svg *Source*: http://en.wikipedia.org/w/index.php?title=File:NYCS-bull-trans-2.svg *License*: unknown *Contributors*: User:Flamurai

File:NYCS-bull-trans-4.svg *Source*: http://en.wikipedia.org/w/index.php?title=File:NYCS-bull-trans-4.svg *License*: unknown *Contributors*: User:Flamurai, User:Svgalbertian

File:NYCS-bull-trans-5.svg *Source*: http://en.wikipedia.org/w/index.php?title=File:NYCS-bull-trans-5.svg *License*: unknown *Contributors*: User:Flamurai

File:NYCS-bull-trans-6.svg *Source*: http://en.wikipedia.org/w/index.php?title=File:NYCS-bull-trans-6.svg *License*: unknown *Contributors*: User:Flamurai, User:Svgalbertian

File:NYCS-bull-trans-7d.svg *Source*: http://en.wikipedia.org/w/index.php?title=File:NYCS-bull-trans-7d.svg *License*: unknown *Contributors*: User:Flamurai

File:NYCS-bull-trans-S.svg *Source*: http://en.wikipedia.org/w/index.php?title=File:NYCS-bull-trans-S.svg *License*: unknown *Contributors*: Andros 1337, Flamurai, Man vyi, Pieter Kuiper, Roomba, Svgalbertian, TLK in 3, Train2104, Wst

File:NYCS-bull-trans-A.svg *Source*: http://en.wikipedia.org/w/index.php?title=File:NYCS-bull-trans-A.svg *License*: unknown *Contributors*: User:Flamurai

File:NYCS-bull-trans-L.svg *Source*: http://en.wikipedia.org/w/index.php?title=File:NYCS-bull-trans-L.svg *License*: unknown *Contributors*: User:Flamurai

File:NYCS-bull-trans-B.svg *Source*: http://en.wikipedia.org/w/index.php?title=File:NYCS-bull-trans-B.svg *License*: unknown *Contributors*: User:Flamurai

File:NYCS-bull-trans-M.svg *Source*: http://en.wikipedia.org/w/index.php?title=File:NYCS-bull-trans-M.svg *License*: unknown *Contributors*: Andros 1337, Flamurai, Man vyi, Roomba, Sarang, Siegele Roland, Svgalbertian, TLK in 3, Train2104, Wst

File:NYCS-bull-trans-C.svg *Source*: http://en.wikipedia.org/w/index.php?title=File:NYCS-bull-trans-C.svg *License*: unknown *Contributors*: User:Flamurai

File:NYCS-bull-trans-N.svg *Source*: http://en.wikipedia.org/w/index.php?title=File:NYCS-bull-trans-N.svg *License*: unknown *Contributors*: User:Flamurai

File:NYCS-bull-trans-D.svg *Source*: http://en.wikipedia.org/w/index.php?title=File:NYCS-bull-trans-D.svg *License*: unknown *Contributors*: User:Flamurai

File:NYCS-bull-trans-Q.svg *Source*: http://en.wikipedia.org/w/index.php?title=File:NYCS-bull-trans-Q.svg *License*: unknown *Contributors*: User:Flamurai

File:NYCS-bull-trans-E.svg *Source*: http://en.wikipedia.org/w/index.php?title=File:NYCS-bull-trans-E.svg *License*: unknown *Contributors*: User:Flamurai

File:NYCS-bull-trans-R.svg *Source*: http://en.wikipedia.org/w/index.php?title=File:NYCS-bull-trans-R.svg *License*: unknown *Contributors*: User:Flamurai

File:NYCS-bull-trans-F.svg *Source*: http://en.wikipedia.org/w/index.php?title=File:NYCS-bull-trans-F.svg *License*: unknown *Contributors*: user:Flamurai

File:NYCS-bull-trans-G.svg *Source*: http://en.wikipedia.org/w/index.php?title=File:NYCS-bull-trans-G.svg *License*: unknown *Contributors*: User:Flamurai

File:NYCS-bull-trans-H.svg *Source*: http://en.wikipedia.org/w/index.php?title=File:NYCS-bull-trans-H.svg *License*: unknown *Contributors*: User:Svgalbertian

File:NYCS-bull-trans-J.svg *Source*: http://en.wikipedia.org/w/index.php?title=File:NYCS-bull-trans-J.svg *License*: unknown *Contributors*: User:Flamurai

File:NYC subway late night map June2010.svg *Source*: http://en.wikipedia.org/w/index.php?title=File:NYC_subway_late_night_map_June2010.svg *License*: unknown *Contributors*: User:CountZ, User:Untogethered

File:Mta station wall.jpg *Source*: http://en.wikipedia.org/w/index.php?title=File:Mta_station_wall.jpg *License*: unknown *Contributors*: User:Dschwen

File:NYC Subway Turnstile Exit Light.jpg *Source*: http://en.wikipedia.org/w/index.php?title=File:NYC_Subway_Turnstile_Exit_Light.jpg *License*: unknown *Contributors*: User:Apricot50

File:59th Street–Columbus Circle (New York City Subway) by David Shankbone.jpg *Source*: http://en.wikipedia.org/w/index.php?title=File:59th_Street–Columbus_Circle_(New_York_City_Subway)_by_David_Shankbone.jpg *License*: unknown *Contributors*: David Shankbone

File:Utica Av IRT EB jeh.JPG *Source*: http://en.wikipedia.org/w/index.php?title=File:Utica_Av_IRT_EB_jeh.JPG *License*: unknown *Contributors*: User:Jim.henderson

File:BwyWalk0505 StationLincolnCenter.jpg *Source*: http://en.wikipedia.org/w/index.php?title=File:BwyWalk0505_StationLincolnCenter.jpg *License*: unknown *Contributors*: User:Aspersions

File:Broadway-Lafayette station by David Shankbone.jpg *Source*: http://en.wikipedia.org/w/index.php?title=File:Broadway-Lafayette_station_by_David_Shankbone.jpg *License*: unknown *Contributors*: David Shankbone

File:Church Av ind toilets jeh.JPG *Source*: http://en.wikipedia.org/w/index.php?title=File:Church_Av_ind_toilets_jeh.JPG *License*: unknown *Contributors*: User:Jim.henderson

File:Astor Place IRT 008.JPG *Source*: http://en.wikipedia.org/w/index.php?title=File:Astor_Place_IRT_008.JPG *License*: unknown *Contributors*: User:Gryffindor

File:New York City subway (2011) IMG 3927.JPG *Source*: http://en.wikipedia.org/w/index.php?title=File:New_York_City_subway_(2011)_IMG_3927.JPG *License*: unknown *Contributors*: User:Billy Hathorn

File:NYC Vaktrak.jpg *Source*: http://en.wikipedia.org/w/index.php?title=File:NYC_Vaktrak.jpg *License*: unknown *Contributors*: User:Dschwen

File:NYC N train cockpit.jpg *Source*: http://en.wikipedia.org/w/index.php?title=File:NYC_N_train_cockpit.jpg *License*: unknown *Contributors*: User:Dschwen

File:Nyc transit authority token.png *Source*: http://en.wikipedia.org/w/index.php?title=File:Nyc_transit_authority_token.png *License*: unknown *Contributors*: User:Jessamyn

File:MetroCard.SVG *Source*: http://en.wikipedia.org/w/index.php?title=File:MetroCard.SVG *License*: unknown *Contributors*: AEMoreira042281, Acps110, Hengsheng120, Sfan00 IMG, Skier Dude, Train2104

File:Fastrack 8 Av vc.jpg *Source*: http://en.wikipedia.org/w/index.php?title=File:Fastrack_8_Av_vc.jpg *License*: unknown *Contributors*: Metropolitan Transportation Authority of the State of New York

File:L train sign.jpeg *Source*: http://en.wikipedia.org/w/index.php?title=File:L_train_sign.jpeg *License*: unknown *Contributors*: Original uploader was Error46146 at en.wikipedia

File:NYC Subway RFID.jpg *Source*: http://en.wikipedia.org/w/index.php?title=File:NYC_Subway_RFID.jpg *License*: unknown *Contributors*: AaronY, DanTD, FlickreviewR, NE2, Vcohen

File:On the go vc.jpg *Source*: http://en.wikipedia.org/w/index.php?title=File:On_the_go_vc.jpg *License*: unknown *Contributors*: Metropolitan Transportation Authority of the State of New York

File:168th Street IRT Broadway 2.JPG *Source*: http://en.wikipedia.org/w/index.php?title=File:168th_Street_IRT_Broadway_2.JPG *License*: unknown *Contributors*: User:Gryffindor

File:NYC subway riders with their newspapers.jpg *Source*: http://en.wikipedia.org/w/index.php?title=File:NYC_subway_riders_with_their_newspapers.jpg *License*: unknown *Contributors*: Travis Ruse, cropped by GallifreyanPostman at en.wikipedia

File:New York City Subway 100 1888 edited.JPG *Source*: http://en.wikipedia.org/w/index.php?title=File:New_York_City_Subway_100_1888_edited.JPG *License*: unknown *Contributors*:

File:Pumping125Street.jpg *Source*: http://en.wikipedia.org/w/index.php?title=File:Pumping125Street.jpg *License*: unknown *Contributors*: Metropolitan Transportation Authority of the State of New York

File:Bowling Green Station Entrance in Flood Prep vc.jpg *Source*: http://en.wikipedia.org/w/index.php?title=File:Bowling_Green_Station_Entrance_in_Flood_Prep_vc.jpg *License*: unknown *Contributors*: Metropolitan Transportation Authority of the State of New York

File:South Ferry Subway Station Entrance under Water vc.jpg *Source*: http://en.wikipedia.org/w/index.php?title=File:South_Ferry_Subway_Station_Entrance_under_Water_vc.jpg *License*: unknown *Contributors*: Metropolitan Transportation Authority of the State of New York

File:Pump Train in Cranberry Street Tunnel after Hurricane Sandy vc.jpg *Source*: http://en.wikipedia.org/w/index.php?title=File:Pump_Train_in_Cranberry_Street_Tunnel_after_Hurricane_Sandy_vc.jpg *License*: unknown *Contributors*: Metropolitan Transportation Authority of the State of New York

File:Service to Far Rockaway (H) vc.jpg *Source*: http://en.wikipedia.org/w/index.php?title=File:Service_to_Far_Rockaway_(H)_vc.jpg *License*: unknown *Contributors*: Metropolitan Transportation Authority of the State of New York

File:Heavily tagged subway car in NY.jpg *Source*: http://en.wikipedia.org/w/index.php?title=File:Heavily_tagged_subway_car_in_NY.jpg *License*: unknown *Contributors*: Calonius, Erik

File:Rats in the NYC Subway 2 vc.jpg *Source*: http://en.wikipedia.org/w/index.php?title=File:Rats_in_the_NYC_Subway_2_vc.jpg *License*: unknown *Contributors*: Paul Lowry from New York, EEUU de A

File:NYC Transit Museum Miss Subways.jpg *Source*: http://en.wikipedia.org/w/index.php?title=File:NYC_Transit_Museum_Miss_Subways.jpg *License*: unknown *Contributors*: Wcnghj

file:NYCS-bull-trans-D.svg *Source*: http://en.wikipedia.org/w/index.php?title=File:NYCS-bull-trans-D.svg *License*: unknown *Contributors*: User:Flamurai

File:Bay Parkway (BMT West End Line).jpg *Source*: http://en.wikipedia.org/w/index.php?title=File:Bay_Parkway_(BMT_West_End_Line).jpg *License*: unknown *Contributors*: DanTD, FlickreviewR, NE2

File:Bmt triplex no 3.png *Source*: http://en.wikipedia.org/w/index.php?title=File:Bmt_triplex_no_3.png *License*: unknown *Contributors*: Cecropia, Kelly

Image:NYCS-SSI-alltimes.svg *Source*: http://en.wikipedia.org/w/index.php?title=File:NYCS-SSI-alltimes.svg *License*: unknown *Contributors*: user:flamurai

file:2nd_ave_subway.svg *Source*: http://en.wikipedia.org/w/index.php?title=File:2nd_ave_subway.svg *License*: unknown *Contributors*: 21:12, 9 July 2008 (UTC)

File:Second Avenue El - demolition.jpg *Source*: http://en.wikipedia.org/w/index.php?title=File:Second_Avenue_El_-_demolition.jpg *License*: unknown *Contributors*: JustSomePics

Image:NYCS-bull-trans-T.svg *Source*: http://en.wikipedia.org/w/index.php?title=File:NYCS-bull-trans-T.svg *License*: unknown *Contributors*: User:Flamurai

Image:NYCS-bull-trans-Q.svg *Source*: http://en.wikipedia.org/w/index.php?title=File:NYCS-bull-trans-Q.svg *License*: unknown *Contributors*: User:Flamurai

File:Second Ave Subway CGI station.jpg *Source*: http://en.wikipedia.org/w/index.php?title=File:Second_Ave_Subway_CGI_station.jpg *License*: unknown *Contributors*: Dream out loud, Utcursch

File:SAS72ndStreet201201.jpg *Source*: http://en.wikipedia.org/w/index.php?title=File:SAS72ndStreet201201.jpg *License*: unknown *Contributors*: Metropolitan Transportation Authority of the State of New York

File:SAS64thStreet201201.jpg *Source*: http://en.wikipedia.org/w/index.php?title=File:SAS64thStreet201201.jpg *License*: unknown *Contributors*: Metropolitan Transportation Authority of the State of New York

File:2nd Av subway dig 83rd St jeh.JPG *Source*: http://en.wikipedia.org/w/index.php?title=File:2nd_Av_subway_dig_83rd_St_jeh.JPG *License*: unknown *Contributors*: User:Jim.henderson

File:SASPhase1TBMBreakthrough.jpg *Source*: http://en.wikipedia.org/w/index.php?title=File:SASPhase1TBMBreakthrough.jpg *License*: unknown *Contributors*: Patrick Cashin, Metropolitan Transportation Authority of the State of New York

File:Yankee Stadium 001.JPG *Source*: http://en.wikipedia.org/w/index.php?title=File:Yankee_Stadium_001.JPG *License*: unknown *Contributors*: User:Gryffindor

File:New York City location Bronx.svg *Source*: http://en.wikipedia.org/w/index.php?title=File:New_York_City_location_Bronx.svg *License*: unknown *Contributors*: Jameslwoodward, NYCRuss, ScottyBerg, The Emirr

File:Bronx 1900.jpg *Source*: http://en.wikipedia.org/w/index.php?title=File:Bronx_1900.jpg *License*: unknown *Contributors*: DanTD, Jim.henderson, Nkabouris

File:Bronx transit and landmarks in 1896 (NY Times) - re-tinted.png *Source*: http://en.wikipedia.org/w/index.php?title=File:Bronx_transit_and_landmarks_in_1896_(NY_Times)_-_re-tinted.png *License*: unknown *Contributors*: The New York Times (no individual author or illustrator credited)

File:Pelhambay1.jpg *Source*: http://en.wikipedia.org/w/index.php?title=File:Pelhambay1.jpg *License*: unknown *Contributors*: Peter J. Romano 2nd

File:Co-op City Hutch River.jpg *Source*: http://en.wikipedia.org/w/index.php?title=File:Co-op_City_Hutch_River.jpg *License*: unknown *Contributors*: Original uploader was Sacme at en.wikipedia

File:CityIsland.jpg *Source*: http://en.wikipedia.org/w/index.php?title=File:CityIsland.jpg *License*: unknown *Contributors*: User:Terryballard

File:Grand Conc 165 jeh.JPG *Source*: http://en.wikipedia.org/w/index.php?title=File:Grand_Conc_165_jeh.JPG *License*: unknown *Contributors*: User:Jim.henderson

File:New Yankee Stadium.JPG *Source*: http://en.wikipedia.org/w/index.php?title=File:New_Yankee_Stadium.JPG *License*: unknown *Contributors*: Original uploader was BuickCenturyDriver at en.wikipedia

File:Bronxhub1.jpeg *Source*: http://en.wikipedia.org/w/index.php?title=File:Bronxhub1.jpeg *License*: unknown *Contributors*: Original uploader was Futurebird at en.wikipedia

File:Aerial view of the Bronx, Harlem River, Harlem, Hudson River, George Washington Bridge, 2008-05-10.jpg *Source*: http://en.wikipedia.org/w/index.php?title=File:Aerial_view_of_the_Bronx,_Harlem_River,_Harlem,_Hudson_River,_George_Washington_Bridge,_2008-05-10.jpg *License*: unknown *Contributors*: Ajay Tallam

File:Fordhamroadbx1.JPG *Source*: http://en.wikipedia.org/w/index.php?title=File:Fordhamroadbx1.JPG *License*: unknown *Contributors*: DanTD, Jim.henderson, T2, Wikiwiki718

File:Bronx-Whitestone Bridge.jpg *Source*: http://en.wikipedia.org/w/index.php?title=File:Bronx-Whitestone_Bridge.jpg *License*: unknown *Contributors*: Dave Frieder

File:Aerial View of the Throgs Neck Bridge.jpg *Source*: http://en.wikipedia.org/w/index.php?title=File:Aerial_View_of_the_Throgs_Neck_Bridge.jpg *License*: unknown *Contributors*: Clindberg, Elipongo, Ingolfson, Ronaldino, Voyager

File:Middletown Road (IRT Pelham Line) by David Shankbone copy.jpg *Source*: http://en.wikipedia.org/w/index.php?title=File:Middletown_Road_(IRT_Pelham_Line)_by_David_Shankbone_copy.jpg *License*: unknown *Contributors*: User:DavidShankbone

File:NYC Transit New Flyer 5400.jpg *Source*: http://en.wikipedia.org/w/index.php?title=File:NYC_Transit_New_Flyer_5400.jpg *License*: unknown *Contributors*: User:AEMoreira042281

File:Bronxpoverty.JPG *Source*: http://en.wikipedia.org/w/index.php?title=File:Bronxpoverty.JPG *License*: unknown *Contributors*: Bender235, NYCRuss, Wikiwiki718

File:BronxScience.jpg *Source*: http://en.wikipedia.org/w/index.php?title=File:BronxScience.jpg *License*: unknown *Contributors*: User:Bxsstudent

File:Fordham University Keating Hall.JPG *Source*: http://en.wikipedia.org/w/index.php?title=File:Fordham_University_Keating_Hall.JPG *License*: unknown *Contributors*: Chriscobar

File:The PLAYERS Club Steppers by David Shankbone.jpg *Source*: http://en.wikipedia.org/w/index.php?title=File:The_PLAYERS_Club_Steppers_by_David_Shankbone.jpg *License*: unknown *Contributors*: David Shankbone

File:Bronx Yankee and park.jpg *Source*: http://en.wikipedia.org/w/index.php?title=File:Bronx_Yankee_and_park.jpg *License*: unknown *Contributors*: Original uploader was The Interloafer at en.wikipedia

File:W125irtjeh.JPG *Source*: http://en.wikipedia.org/w/index.php?title=File:W125irtjeh.JPG *License*: unknown *Contributors*: User:Jim.henderson

File:125th Street Harlem.jpg *Source*: http://en.wikipedia.org/w/index.php?title=File:125th_Street_Harlem.jpg *License*: unknown *Contributors*: User:dananderson

File:Harlem - W125st - Madison Avenue.jpg *Source*: http://en.wikipedia.org/w/index.php?title=File:Harlem_-_W125st_-_Madison_Avenue.jpg *License*: unknown *Contributors*: Spiritia, 1 anonymous edits

File:55w125.jpg *Source*: http://en.wikipedia.org/w/index.php?title=File:55w125.jpg *License*: unknown *Contributors*: User:Americasroof

File:Qns blvd 39st jeh.JPG *Source*: http://en.wikipedia.org/w/index.php?title=File:Qns_blvd_39st_jeh.JPG *License*: unknown *Contributors*: User:Jim.henderson

Image:NYCS-SSI-weekdaysonly.svg *Source*: http://en.wikipedia.org/w/index.php?title=File:NYCS-SSI-weekdaysonly.svg *License*: unknown *Contributors*: user:flamurai

Image:NYCS-SSI-allexceptnights.svg *Source*: http://en.wikipedia.org/w/index.php?title=File:NYCS-SSI-allexceptnights.svg *License*: unknown *Contributors*: user:flamurai

Image:NYCS-SSI-rushpeak.svg *Source*: http://en.wikipedia.org/w/index.php?title=File:NYCS-SSI-rushpeak.svg *License*: unknown *Contributors*: user:flamurai

Image:NYCS-SSI-rushonly.svg *Source*: http://en.wikipedia.org/w/index.php?title=File:NYCS-SSI-rushonly.svg *License*: unknown *Contributors*: user:flamurai

Image:NYCS-SSI-nightsonly.svg *Source*: http://en.wikipedia.org/w/index.php?title=File:NYCS-SSI-nightsonly.svg *License*: unknown *Contributors*: user:flamurai

File:R62A Showing 12 Sign.jpg *Source*: http://en.wikipedia.org/w/index.php?title=File:R62A_Showing_12_Sign.jpg *License*: unknown *Contributors*: User:R36

Image:Second Avenue in New York by David Shankbone.jpg *Source*: http://en.wikipedia.org/w/index.php?title=File:Second_Avenue_in_New_York_by_David_Shankbone.jpg *License*: unknown *Contributors*: David Shankbone

File:Second Avenue Manhattan 1861.jpg *Source*: http://en.wikipedia.org/w/index.php?title=File:Second_Avenue_Manhattan_1861.jpg *License*: unknown *Contributors*: Not identified

File:Grand Street Station Entrance.jpg *Source*: http://en.wikipedia.org/w/index.php?title=File:Grand_Street_Station_Entrance.jpg *License*: unknown *Contributors*:

File:239th Street Yard, NYC IRT Subway, Summer, 2010.jpg *Source*: http://en.wikipedia.org/w/index.php?title=File:239th_Street_Yard,_NYC_IRT_Subway,_Summer,_2010.jpg *License*: unknown *Contributors*: DanTD, DragonflySixtyseven, Future Perfect at Sunrise, Jim.henderson, SConnick

File:239th St Yard Bx jeh.JPG *Source*: http://en.wikipedia.org/w/index.php?title=File:239th_St_Yard_Bx_jeh.JPG *License*: unknown *Contributors*: User:Jim.henderson

File:R110a r62a at 239th street yard.JPG *Source*: http://en.wikipedia.org/w/index.php?title=File:R110a_r62a_at_239th_street_yard.JPG *License*: unknown *Contributors*: User:bebo2good1

Image:Waldo Av 1 train yard jeh.JPG *Source*: http://en.wikipedia.org/w/index.php?title=File:Waldo_Av_1_train_yard_jeh.JPG *License*: unknown *Contributors*: User:Jim.henderson

File:Corona Yard jeh.JPG *Source*: http://en.wikipedia.org/w/index.php?title=File:Corona_Yard_jeh.JPG *License*: unknown *Contributors*: User:Jim.henderson

File:Corona train yard.jpeg *Source*: http://en.wikipedia.org/w/index.php?title=File:Corona_train_yard.jpeg *License*: unknown *Contributors*: Original uploader was Error46146 at en.wikipedia

Image:IRT E180-UnionP Yard.JPG *Source*: http://en.wikipedia.org/w/index.php?title=File:IRT_E180-UnionP_Yard.JPG *License*: unknown *Contributors*: User:bebo2good1

Image:NYCS IRT shop 1902.jpg *Source*: http://en.wikipedia.org/w/index.php?title=File:NYCS_IRT_shop_1902.jpg *License*: unknown *Contributors*: Interborough Rapid Transit Company

File:Livoniayardjeh.JPG *Source*: http://en.wikipedia.org/w/index.php?title=File:Livoniayardjeh.JPG *License*: unknown *Contributors*: User:Jim.henderson

File:WTM3 The Fixers 0027.jpg *Source*: http://en.wikipedia.org/w/index.php?title=File:WTM3_The_Fixers_0027.jpg *License*: unknown *Contributors*: The Fixers (Wikis Take Manhattan 2009 participant)

File:NYCS 207th St Yard.jpg *Source*: http://en.wikipedia.org/w/index.php?title=File:NYCS_207th_St_Yard.jpg *License*: unknown *Contributors*: User:Doodle77

File:CI Yard Entrance jeh.jpg *Source*: http://en.wikipedia.org/w/index.php?title=File:CI_Yard_Entrance_jeh.jpg *License*: unknown *Contributors*: User:Jim.henderson

File:Coney Island Complex Outside.JPG *Source*: http://en.wikipedia.org/w/index.php?title=File:Coney_Island_Complex_Outside.JPG *License*: unknown *Contributors*: User:Wusel007

File:Coney Island Complex Workshop.JPG *Source*: http://en.wikipedia.org/w/index.php?title=File:Coney_Island_Complex_Workshop.JPG *License*: unknown *Contributors*: User:Wusel007

file:USA_New_York_City_location_map.svg *Source*: http://en.wikipedia.org/w/index.php?title=File:USA_New_York_City_location_map.svg *License*: unknown *Contributors*: User:Alexrk2

File:Red pog.svg *Source*: http://en.wikipedia.org/w/index.php?title=File:Red_pog.svg *License*: unknown *Contributors*: Anomie

File:NYCS Concourse yard.jpg *Source*: http://en.wikipedia.org/w/index.php?title=File:NYCS_Concourse_yard.jpg *License*: unknown *Contributors*: Flowizm / ...some guy

File:Concourseydjeh.jpg *Source*: http://en.wikipedia.org/w/index.php?title=File:Concourseydjeh.jpg *License*: unknown *Contributors*: User:Jim.henderson

File:Jamaica Yard east jeh.JPG *Source*: http://en.wikipedia.org/w/index.php?title=File:Jamaica_Yard_east_jeh.JPG *License*: unknown *Contributors*: User:Jim.henderson

File:Jamaica Yard doorway jeh.JPG *Source*: http://en.wikipedia.org/w/index.php?title=File:Jamaica_Yard_doorway_jeh.JPG *License*: unknown *Contributors*: User:Jim.henderson

File:Pitkin Av Yard jeh.JPG *Source*: http://en.wikipedia.org/w/index.php?title=File:Pitkin_Av_Yard_jeh.JPG *License*: unknown *Contributors*: User:Jim.henderson

File:Rockaway Yard east jeh.jpg *Source*: http://en.wikipedia.org/w/index.php?title=File:Rockaway_Yard_east_jeh.jpg *License*: unknown *Contributors*: User:Jim.henderson

File:Rockway Yard west jeh.jpg *Source*: http://en.wikipedia.org/w/index.php?title=File:Rockway_Yard_west_jeh.jpg *License*: unknown *Contributors*: User:Jim.henderson

File:BMT 36-38 St yard jeh.JPG *Source*: http://en.wikipedia.org/w/index.php?title=File:BMT_36-38_St_yard_jeh.JPG *License*: unknown *Contributors*: User:Jim.henderson

MIX
Papier aus verantwortungsvollen Quellen
Paper from responsible sources
FSC® C105338

Printed by Books on Demand GmbH, Norderstedt / Germany